# modern oilhydraulic engineering

by Jean U. Thoma

Dr. Sc., Phys.dipl. EPF, A.M.I. Mar. E.
Visiting Professor of the University of Maryland, USA,
and Consultant.

This book is a revised edition of

'Grundlagen der Oelhydraulik'

published by Carl Hanser Verlag, Munich, Germany, and of

'Ingeneria oleodinamica'

published by Casa editrice  Tecniche Nuove, Milano, Italy.


Author's permanent address : Bellevue 23, CH 6300 Zug, Switzerland.


© TRADE & TECHNICAL PRESS LTD.,

85461 043 X


PRINTED AND PUBLISHED BY TRADE & TECHNICAL PRESS LTD.,

# PREFACE

This book treats the design and application of oil hydraulic components and especially hydrostatic machines with fixed or adjustable displacement. It is based on the principles of fluid mechanics which are developed in the first chapters in a form suitable for the considerable viscosity of hydraulic oils. Especially important are the conditions of oil flow and pressure distribution in narrow gaps which are treated in Chapter 2 with the relationship to full film lubrication in general mechanical engineering. After a chapter on valves, the rest of the book is devoted to the fundamentals of hydraulic circuits, to practical aspects of servocontrol design, to hydrostatic transmissions and, finally, to vibrations and shock waves.

The book is based on a course of the author at the Engineering School of Constance, Germany, and on lectures and courses held by the author at different American universities. Furthermore, he has added some material of interest to designers and to the industry. As this is not needed for the comprehension of the text, it is contained in small print or in footnotes.

Also in footnotes, the author has included some amusing remarks from the lectures. Apart from bringing a little humour into the life of the engineer, they serve the evident pedagogic purpose of helping to understand and to remember some important concepts.

In spite of their importance, hydrostatic transmissions are treated here only very briefly, since they are the subject of another book by the author of the same publisher. In fact, a satisfactory treatment of circuits, of experiments, of control and of application problems of hydrostatic transmissions requires so much space that a separate volume is well justified.

This book contains only very simple mathematical expressions. Functional diagrams are largely used for representing the interaction and relation of the variables in oil hydraulic components. This powerful method was used extensively in the book 'Hydrostatic Power Transmission' and is described here in Appendix 2.

Several figures were contributed by firms of several countries for which the author is grateful. Such figures are acknowledged in the caption by a shortened name of the contributor whilst a list of the full names and addresses of the contributing firms appears at the end of the book.

The author hopes that this book will prove useful for the practice and the progress of this special discipline :

**Oil Hydraulic Engineering**

Jean U. Thoma

**Zug, Switzerland, and**
   **College Park, Maryland, USA,**


**Spring 1970**

# CONTENTS

# NOTATION

In order to obtain an easily memorised notation, the variables of a certain dimension are denoted as far as possible by the same letter. Variables of the same dimension are distinguished by indices and the preferred units have been added in square brackets.

| | |
|---|---|
| $a$ | velocity of shock waves in conduits [cm/sec] |
| $b$ | width, especially of sealing lips [cm] |
| $c_f$ | dimensional constant in 11–5 |
| $c_v$ | specific heat of oil |
| $c_f \; c_h \; c_v \; c_s$ | loss coefficients in mathematical models in 31 |
| $d$ | diameter [cm] |
| $d_e \; d_m \; d_i$ | external, mean and internal diameter of slipper |
| $d_p \; d_t$ | diameter of piston, of pitch circle in 34 |
| $e$ | eccentricity in 34 [cm] |
| $e$ | wall thickness in 72 [cm] |
| $f$ | function or transfer function of a block |
| $f_{in}$ | transfer function of the internal loop in 54 |
| $g$ | acceleration of gravity |
| $g$ | transfer function of a block in A3 |
| $h$ | height, especially gap height in Chapter 2 [cm] |
| $h_1 \; h_o$ | height at the beginning and end of gap |

| | |
|---|---|
| $i$ | electric control current of servovalves |
| $k$ | spring constant |
| $m$ | mass [kg] |
| $m$ | diametral pitch in 34 [cm] |
| $p$ | pressure [bar] |
| $p_L$ | loss pressure defined by equation in 31–6 |
| $p_{Lf}\ p_{Lh}\ p_{Lv}$ | components of loss pressure in mathematical models in 31 |
| $q$ | displacement per revolution [cm³/rev] |
| $\bar{q}$ | displacement per radiant [cm³/rd] |
| $\bar{q}_o$ | maximum displacement per radiant |
| $\bar{q}_{10}\ \ \bar{q}_{20}$ | maximum displacement per radiant of primary or secondary |
| $s$ | stroke or linear position [cm] |
| $s$ | Laplace variable in 54 |
| $s_c$ | tooth tip clearance in 34 |
| $s_{Ap},\ s_E,\ s_{sv}$ | position of actuating piston, of input, of spool valve in 54 |
| $s_p$ | position of piston in 32 |
| $t$ | time [sec] |
| $u$ | integration variable in Chapter 2 |
| $v$ | velocity or speed [cm] |
| $v_o\ \ v_m$ | wall and mean velocity |
| $x$ | co-ordinate in direction of gap height |
| $y$ | co-ordinate in direction of gap width |
| $z$ | co-ordinate in direction of gap length |
| $z$ | number of cylinders or teeth |
| $A$ | area [cm²] |
| $A_t$ | cross-section of a tooth in 34 |
| $B$ | compressibility modulus defined by equation 12–2 [bar] |
| $D$ | operator of time derivation in Chapter 5 |
| $E$ | modulus of elasticity (Young's modulus) [bar] |
| $F$ | force [daN or kN] |
| $J$ | heat equivalent |

| | |
|---|---|
| $K$ | gain between spool position and flow of a servovalve $[cm^2/sec]$ |
| $L$ | length $[cm]$ |
| $L_H$ | stroke |
| $M$ | torque or turning moment $[daNcm]$ |
| $P$ | power $[daNcm/sec$ or $kW]$ |
| $P_{ap}$ | apparent power in 32 and 63 |
| $Q$ | volume flow $[cm^3/sec]$ |
| $R$ | radius $[cm]$ |
| $T$ | temperature |
| $V$ | volume $[cm^3]$ |
| $V_{cc}$ | volume of compression chambers in 33 |
| $Y$ | admittance $[cm^3/bar\ sec]$ |
| $Y_s$ | leakage admittance |
| $Y_{sc}\ Y_{tg}$ | secant and tangent admittances |
| $Z$ | impedance $[bar\ sec/cm^3]$ |
| $Z_{sc}\ Z_{tg}$ | secant and tangent impedances |
| $\alpha$ | displacement setting |
| $\hat{\alpha}$ | angle of inclination or tilt |
| $\beta$ | balancing factor in underbalanced hydrostatic bearings |
| $\delta$ | ratio of gap heights in 23 |
| $\delta$ | wall thickness as fraction of piston diameter |
| $\gamma$ | lap angle of valve faces in 33 |
| $\gamma\ \hat{cr}$ | angle of connecting rods in 33 |
| $\gamma$ | torque multiplication in 63 |
| $\epsilon$ | strain in 73 |
| $\zeta$ | loss coefficient |
| $\zeta$ | damping coefficient in 54 |
| $\eta$ | efficiency |
| $\eta_m,\ \eta_v$ | mechanical, volumetric and total efficiencies |
| $\eta_{tot}$ | |
| $\theta$ | jet angle |
| $\kappa$ | constant for gas accumulators |
| $\lambda$ | load factor of hydrostatic bearings |

x

| | |
|---|---|
| $\lambda$ | length factor for model laws in 35 |
| $\lambda_R$ | friction factor in tubes |
| $\mu$ | viscosity [barsec] |
| $\rho$ | mass density kg/cm$^3$ |
| $\sigma$ | elastic stress [bar] |
| $\tau$ | shear stress [bar] |
| $\phi$ | angle of rotation [rad] |
| $\omega$ | rotation frequency [rd/sec] |
| $\omega_1, \omega_2, \omega_n$ | rotation frequency of primary, of secondary and of output shaft in Chapter 6 |
| $\omega_n$ | resonance frequency in 54 |
| $\omega_s$ | slip frequency defined by equation 31—5 |
| $\Omega$ | specific rotation frequency of hydrostatic machines defined by equation 35—1 |

# HISTORICAL INTRODUCTION

This book is concerned with the design principles of the many products that constitute today the discipline of oil hydraulic or fluid power engineering. This discipline is quite different from the hydraulics of rivers and dams where the gravity forces are of preponderant importance.

Oil hydraulic engineering is based on the transmission of forces and power by the static pressure of fluids in tubes and conduits as already indicated by Pascal in the 17th century. The first industrial application was developed by Joseph Bramah (1749–1814) in London in the form of a hydraulic press producing large forces with water as working fluid. The most important detail progress was the development of suitable packings and seals between the press piston and the cylinder. Bramah manufactured in his own works presses for compacting wool and other textiles, for pressing oil out of plants and also for tearing out trees. This proceeding was more efficient in wood production, already important at that time since there was a shortage of oak wood suitable for ship building. /FN$^1$/.

Fig 1 shows a schematic presentation of a hydrostatic force transmission on a fluid column between the small piston of the pump and the large press piston. According to Pascal's principle, the fluid column must be contained or supported laterally by the walls of the conduit in contrast to steel columns. This arrangement forms the

FN$^1$ — Information taken from the book : Ian McNeil, 'Joseph Bramah', David and Charles, 1968.

basis not only for the first application but also for modern hydrostatic engineering where suitable control valves control the fluid and its flow and where the handpump is replaced by hydrostatic machines.

In the second half of the 19th century, many hydrostatic machines and components were developed by W.G.Armstrong and used principally for marine engineering for anchor winches and lifting gears. Armstrong already used weight-loaded accumulators in order to equalise the consumption of pressure water. Also, rotary output radial piston machines with valve faces made by pairing of metal and of a heavy tropical wood, the so-called lignum vitae, were used. /FN$^2$/

The development of electric machines and drives resulted in a certain stagnation of hydrostatic engineering until, in 1905, for the first time oil was used as a working fluid. A hydrostatic transmission with axial piston machines was designed by Janney for the training of gun turrets on battle-ships. The main progress was the lubricating effect of the oil as working fluid.

Speed governors and actuators for the large valves of water turbines, introduced about 1910, are another important application of oil hydraulic engineering.

Oil hydraulic radial piston machines were introduced in 1910 by Hele Shaw and in 1922 by Hans Thoma. The latter also developed the tilting head axial piston machine, first (in 1930) with plane valve faces and synchronisation by universal joint, later (in 1946) with spherical valve faces and synchronisation by connecting rods. The swashplate machine has been used industrially since about 1950.

A further important development were the valves, especially the dipole valves including the piloted pressure relief valves, introduced in 1936 by Harry Vickers. Another notable component is the accumulator with gas as elastic medium and with a rubber membrane for the separation of the oil and the gas developed since 1950 by Jean Mercier.

Further research and development were carried out by Blackburn, Lee and others at the American Technical University MIT on hydrostatics under high pressures and servocontrols. They resulted, in about 1958, in the electrohydraulic servovalve applied today almost universally.

FN$^2$ — Many of these interesting old designs are exhibited in the Science Museum, Kensington, London.

These developments gave all necessary components to hydrostatic engineering and led, since 1950, to a large expansion in many fields of application, especially in connection with the trend to automation and rationalisation. The newest tendency is the combination of oil hydraulic components with electronic and fluidic parts. Such systems offer many interesting solutions including the feed drive for numerically controlled machine tools. A proverb known for many years is today more applicable than ever : *'Electrical nerve, hydraulic muscle.'*

It remains as a question for the future how much of the electrical 'nerves' or information handling parts will be replaced by fluidic components with or without moving solids such as membranes.

# 1  FLUID MECHANICS FOR HYDROSTATIC SYSTEMS

Hydrostatic machines and components work with fluids under pressure and their design depends on the laws of fluid mechanics. However, compared to conventional fluid mechanics, the point of view adapted in this book is somewhat different. Therefore, it seems justifiable to repeat something from fluid mechanics in this chapter.

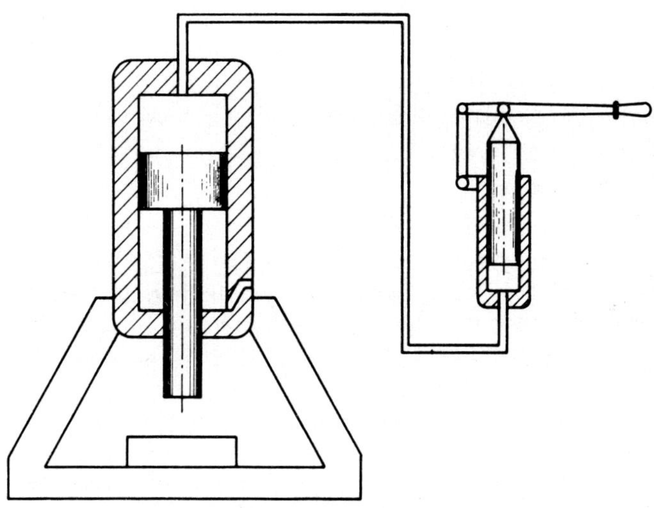

**Fig 1.1: Principle of force transmission between two pistons by fluid columns**

## 11    Fundamentals

### 11-1    Forces in Fluids

Contrary to solids, a fluid cannot sustain shear stress without motion. Therefore, the stresses inside the fluid are equal in all directions and are called static or hydrostatic pressure. /FN[1]/ The static pressure is, therefore, acting equally in all directions according to the principles of Pascal discovered in the 17th century.

A further important property of the fluid is that it is almost incompressible. This means that a given quantity of fluid reduces its volume only very little if an external pressure is applied, as we shall see in detail in section 12. This property is quite opposed to gases which are strongly compressible and which give many new problems in the sister technique 'pneumatics'.

Hydrostatic machines are built for incompressible fluids. Furthermore, the fluid itself can be used for lubrication so that a separate circuit for the lubricant, as in automobile engines, is unnecessary.

In moving or stationary fluids, there are the following force effects:
1. Forces due to pressure.
2. Shear forces due to viscosity.
3. Acceleration forces.
4. Gravity forces.
5. Forces due to surface tension.

The entire conventional fluid mechanics is built from the study of the above force effects, mostly by establishing an equilibrium of forces within a determined volume. Additionally, one has the condition of the conservation of mass, the so-called equation of continuity.

The characteristics of oil hydraulics are that the force effects due to static pressure and due to viscosity are preponderant, whilst the inertia and the gravity forces are practically only perturbations and the surface tensions almost always negligible.

---

FN[1] — We exclude here the so-called non-Newtonian fluids which have no importance in hydrostatic engineering.

The ratio of the forces is indicated in fluid mechanics by the following dimensionless expressions or numbers:

1. The Reynolds number gives the ratio of the inertia to the viscosity forces and is very frequently used. In hydrostatics, however, it is not so useful.

2. The Froude number gives the relation of inertia to gravity forces and is especially important for flow in rivers and channels, and with surface waves on seas.

3. The Weber number gives the ratio between inertia forces and the effects of surface tension.

## 11—2    Basic Hydrostatic Variables

The study of hydrostatics and the design of hydrostatic machines are governed by the action of the basic hydrostatic variables. Therefore, the most important of these variables will be explained here.

Following G. Falk /$FN^2$/ we consider the variables as a basic concept of physics. We shall use the expression 'variable' instead of quantity' even if they are constant. The magnitude of a variable will be its instantaneous value and a constant is a variable having always the same magnitude as, for instance, the acceleration due to gravity.

Built from the basic physical variables length L, time t, force F, mass m, the following variables are especially important in hydrostatics:

The pressure, designated by p, is the ratio of force and area. Conversely, the forces exerted on an area, for instance on the wall of an oil tank, are given by the product of the area and of the pressure.

The volume flow or oil flow /$FN^3$/ is the ratio of the volume flowing through a control area into a suitable measuring device, for instance into a flow-meter, to the necessary time.

The mass flow, i.e. the ratio of the mass flowing through a control area and the necessary time, is less important in hydrostatics. It is given by the product of volume flow and of mass density.

Volume flow and pressure are the most important variables of hydrostatics and anologous to current and voltage in electric engineering. However, according to the experiences of the author, it is not advisable to develop the analogies between electrical and hydrostatic systems too far because the properties of the variables are considerably different in details.

$FN^2$ — G. Falk, 'Theoretische Physik', Springer, 1968.

$FN^3$ — In American literature frequently but improperly called 'flow rate'

We use the example of a simple piston and cylinder in fig 1 to illustrate the relation between hydrostatic and mechanical variables. The flow Q and the pressure p act from the left on the cylinder, whilst the piston with the area A delivers to the right the force F and velocity b.

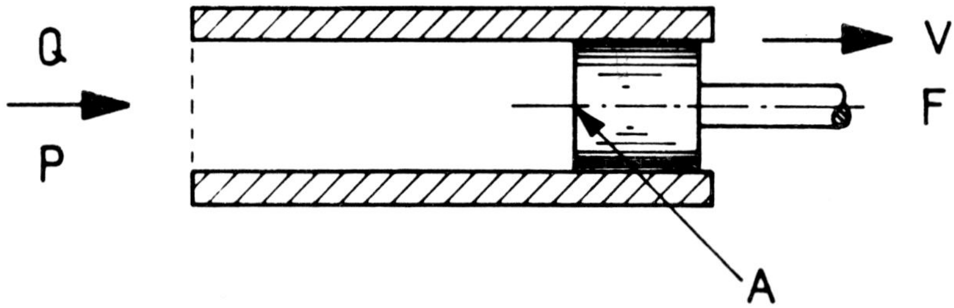


**Fig 11.1:** Hydraulic cylinder with piston as an example for the relation between mechanical and hydrostatic variables

We know from the basic properties of liquids that the force and the pressure are connected as follows

$$F = A\,p \tag{11-1}$$

and the flow and the velocity

$$v = Q/A \tag{11-2}$$

The mechanical power is defined as the product of force and velocity or of torque and rotation frequency. /FN[4]/

$$P_{me} = F\,v = M\,\omega \tag{11-3}$$

The hydrostatic power is equal to the mechanical power if losses are neglected. Inserting of equation 1 and 2 into equation 3 gives

$$P_{he} = Q\,p \tag{11-4}$$

FN[4] — We use the modern expression 'rotation frequency' instead of the more usual word 'rotative speed' or rpm since it is effectively a variable of the dimension of a frequency.

The above relations are strictly valid only for the transport of power by displacement of a completely incompressible and massless fluid against an opposing pressure. Due to the finite compressibility, there is always some energy contained in the fluid itself. In principle, it can be recuperated since the piston continues to travel a little after closing down the flow with progressively decreasing force. Furthermore, the volume flow with given mass flow depends on the static pressure.

The above effects have little importance in oil hydraulics, whilst they play a great role in pneumatics.

## 11—3   Systems of Units and Practical Computations

This book is entirely based on the S.I. or International System of measuring units which considerably facilitates many calculations. For the convenience of the reader, we shall treat briefly the question of computation in the S.I. system.

For the purposes of engineering, it is advisable to consider the following four fundamental variables, from which the entire mechanics is built:

| | |
|---|---|
| Length | L |
| time | t |
| force | F |
| mass | m |

Force and mass of an accelerated object are connected by the Second Law of Newton. It states that the acceleration force of a body is proportional to its mass which can be written as follows

$$F = c_f \, m \, a \tag{11—5}$$

where :   $c_f$ = dimensional constant, m = mass,

a = acceleration

We, therefore, do not use the usual definition according to which force is explained as the product of mass and acceleration. This is entirely possible, but sometimes it is easier to understand engineering without it. Rather, we consider force as a physical fundamental variable as known from the daily experience and from elementary physics. This attitude facilitates understanding of much of engineering.

It is naturally possible to relate the mass of an object and the forces as physical quantities. However, in this case, understanding becomes easier if one considers force and not mass as fundamental variable, in addition to the usual fundamental variables of length and time. Then, the mass of an object becomes a derived variable and gives the relation between its acceleration and an applied force supposing that other forces like friction are negligible. This becomes still clearer if one defines the mass as ratio of the momentum (given by the time integral of the applied force) and the velocity of a body.

$$m = \frac{\int F\, dt}{v}$$

See also R. Fleischmann, 'Die Struktur des physikalischen Begriffsystems', Zeitschrift für Physik, Band 129, pages 3 and 7 – 400, (1951).

The force as variable is more fundamental than the mass since it indicates an energy which is converted by displacing an object by a given length. This energy is called more precisely 'displacement energy'. The mass, on the other hand, is a constant variable which indicates the resistance or inertia of an object against acceleration, as described by equation 5.

The constant $c_f$ of equation 4 depends on the units of measurement.

With proper choice of the units, the constant becomes equal to one. This is the case in the S.I. system of units where the fundamental variables are expressed in the following units :

> Length in metres,
> time in seconds,
> force in newtons,
> mass in kilograms.

The units of the derived quantities are products or ratios of the fundamental quantities as, for instance, m/sec for velocities and m/sec$^2$ for accelerations.

The newton is, therefore, the force required to accelerate an object of the mass of one kilogram with one m/sec$^2$.

Gravity acts on all objects like an acceleration, the so-called acceleration of gravity ($g = 9.81$ m/sec$^2$ = 32.3 ft/sec$^2$). If an object remains at rest In spite of this, it applies a force on its support. This is the so-called weight force or weight given by equation 5. In the S.I. system, the weight of an object with one kilogram mass is equal to 9.81 newtons, from which any flotation force has to be deducted according to the principle of Archimedes.

Some units of measurement have the disadvantage that they become very small or very large. Therefore, one uses suitable prefixes in order to indicate powers of ten of the units of measurement. These prefixes can be applied to any physical quantity and we can cite as example kilogram, millimetre or microsecond.

For completeness, we still mention the old technical system of units in which the unit force is the kilogram force defined as the weight of an object of one kilogram mass at normal conditions on the surface of the earth. This means that the constant c in equation 5 is not equal to 1 but equal to 1/9.81. This is correct in principle, only one must not forget for high accuracy that the actual acceleration of gravity depends slightly on latitude and on altitude and that, furthermore, the actual weight is influenced by flotation of the object in the air.

One obtains therefore  1 kgf = 9.81 N  ≈ 1 daN

We shall use the following units of measurement in this book :

Force : The dekanewton which corresponds to a kilogram force (exactly 1 daN = 1.02 kgf) and the kilonewton which is a very convenient with larger forces, as, for instance, with the load capacity of roller bearings.

Pressure : The bar (1 bar = $10^5$ N/m$^2$ = 1 daN/cm$^2$) corresponding to 1 atmosphere and the hectobar :   1 hebar = $10^8$   N/m$^2$ = 1 kN/cm$^2$ ≈  1 kgf/mm$^2$.

The latter unit is especially convenient for strength calculation of materials because there it is usual to express the admissible elastic tensions in kgf/mm$^2$.

Power : The watt (1 W = 1 Nm/sec) and the kilowatt (1 kW = 1 kNm/sec). The horse power is no longer used. /FN$^5$/

Torque : In dekanewtonmeters (daNm) equal to one kilonewtoncentimeter (kNcm).

Rotation frequency : /FN$^6$/ Starting from the usual indication in revolutions per minute, we prefer to use radiants per second denoted by $\omega$. Since

---

FN$^5$ — Amusing remark from the lectures : 'The horse has served us well since the time of the first steam engine for the measurement of power. Let it now retire in peace.

FN$^6$ — Sometimes called rotative speed.

1  radiant $= 1/2\pi$  revolution $= 57°$, one obtains

$$\frac{\omega}{\text{rd/sec}} = \frac{6.28}{60} \cdot \frac{n}{\text{rpm}} = 0.105 \cdot \frac{n}{\text{rpm}}$$

that is about 10% of the value obtained when using the unit revolutions per minute. Table 1 contains the conversions of some usual values.

**Table 11 1**

Conversion of rotation frequencies from rd/sec in rpm

| ( rd/sec) | 104 | 157 | 208 | 314 | 628 |
|---|---|---|---|---|---|
| ( rpm   ) | 1 000 | 1 500 | 2 000 | 3 000 | 6 000 |

Volume flow : In $cm^3$/sec and in lit/sec = 1 000 $cm^3$/sec. In practice, one also frequently uses litres per minute where one has 60 lit/min = 1 000 $cm^3$/sec.

This system is a slight adaption of the S.I. system for the ease of practical computation since only powers of ten appear as conversion factors. /FN[7]/ We would only like to remind that, for converting a quantity to a ten times smaller unit of measurement, the measure i.e. the numerical value must be multiplied by ten.

### Numerical Example

Let us calculate as an exercise the piston of a small hydraulic press. What force is developed by a piston of 80 mm dia under a pressure of 120 bar and what flow is required for a speed of 20 cm/sec?

### Solution

With a piston area of 50 $cm^2$, one computes the force from equation 1

$$F = 50 \text{ cm}^2 \cdot 120 \frac{\text{daN}}{\text{cm}^2} = 6\,000 \text{ daN} = 60 \text{ kN}$$

FN[7] — Such conversion factors can never be entirely avoided as long as lengths are indicated in millimeters and not in meters on engineering drawings.

and the flow from equation 2

$$Q = A \cdot v = 50 \text{ cm}^2 \cdot 20 \frac{\text{cm}}{\text{sec}} = 1\,000 \text{ cm}^3/\text{sec}$$

The converted mechanical power becomes according to equation 3

$$P_{me} = 60 \text{ kN} \cdot 20 \frac{\text{cm}}{\text{sec}} = 1\,200 \text{ kN} \frac{\text{cm}}{\text{sec}} = 12 \text{ kW}$$

and the hydraulic power

$$P_{he} = 1\,000 \frac{\text{cm}^3}{\text{sec}} \quad 120 \frac{\text{daN}}{\text{cm}^2\text{sec}} = 120\,000 \frac{\text{daNcm}}{\text{sec}} = 12 \text{ kW}$$

## 11–4    Forces on Walls

Hydrostatic pressure in liquids affects the walls of the vessel or container and produces there a force as given as equation 1 above. In fig. 1. the pressure acts not only on the piston but also on the walls of the cylinder or, in general, it presses the walls of vessels outward. It is the most important effect in oil hydraulics.

According to the principle of Pascal, there should be equal pressure in all parts of the liquid. In practice, some deviations are due to the following three effects :

1. Due to the weight of the liquid, the pressure increases, for instance in a tank, with the vertical distance from the free surface.

2. If liquid is flowing through narrow gaps, it experiences a considerable pressure drop. Therefore, a variable local pressure acts on the walls of the gap.

3. If the liquid is suddenly accelerated, for instance at a restriction of the flow area in a nozzle, the pressure decreases because it has to supply the accelerating force (see Section 13–1)

The effect 3 is of practical importance mostly in connection with cavitation as treated in section 12. On the other hand, effect 2 influences the flow in gaps to which we shall devote the chapter 2. Effect 1 is interesting in principle but important in practice only with oil tanks.

With variable pressure, the total force on the wall is obtained by integration of the pressure field over its surface, in place of equation 1

$$F = \int p \, d A \tag{11--6}$$

In order to calculate the pressure development under the influence of the weight of the liquid, consider a vertical cylinder as shown in fig 2. The cylinder is closed at the bottom by a piston of area A and filled to the height h with oil of mass density $\rho$. The force on the piston due to the weight is equal to the product of the mass of the liquid and the acceleration of gravity g.

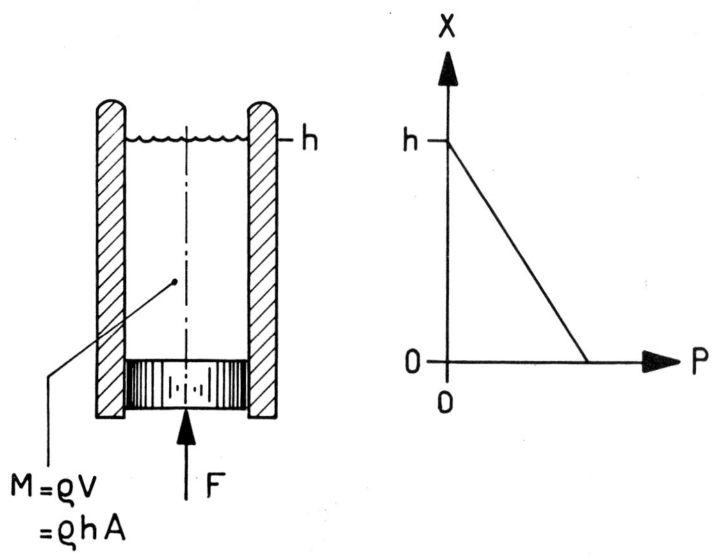


Fig. 11.2: Gravity forces and pressure distribution in
a vertical cylinder

$$F = A p = \rho A h g \tag{11--7}$$

Within the oil column, only the mass situated above the point x contributes to hydrostatic pressure. Therefore, one obtains for the pressure

$$p = \rho g(h - x) \tag{11--8}$$

The pressure as function of the height according to equation 8 is represented on the right-hand side of fig 2. The entire force tending to push the wall of the cylinder outward is obtained by integration of the pressure over the surface of the wall.

### Numerical Example for the Weight of Liquid

What is the pressure due to gravity in an oil tank 70 cm (28 in) below the surface?

Mass density of oil $\rho = 900 \text{ kg/m}^3 = 900 \cdot 10^{-6} \text{ kg/cm}^3$

According to formula 8 with $h - x = 70$ cm and $g = 981 \text{ cm/sec}^2$

$$p = 900 \cdot 10^{-6}\,\frac{\text{kg}}{\text{cm}^3} \cdot .981\,\frac{\text{cm}}{\text{sec}^2} \cdot 70\,\text{cm} = 62\,\frac{\text{kg}}{\text{cmsec}^2}$$

Now

$$1\,\text{kg}\,\frac{\text{m}}{\text{sec}^2} = 1\,\text{N, thus}\quad 1\,\text{kg}\,\frac{\text{cm}}{\text{sec}^2} = 10^{-2}\,\text{N and}$$

$$1\,\frac{\text{kg}}{\text{cmsec}^2} = 10^{-2}\,\frac{\text{N}}{\text{cm}^2} = 10^{-3}\,\frac{\text{daN}}{\text{cm}^2} = 1\,\text{mbar (millibar)}$$

We obtain thus

$$p = 62\,\text{mbar}$$

Since the pressure is proportional to the vertical distance, we obtain with 100 cm (39 in) a pressure of 90 mbar (1.3 psi). For practical calculations, it is convenient to express this as follows

$$\frac{p}{90\,\text{mbar}} = \frac{h}{100\,\text{cm}} \tag{11—8a}$$

or in words:

The pressure in an oil column with a free surface is as many times 90 mbar (1.3 psi) as the vertical distance a multiple of 100 cm (39 in).

## 12    Properties of Hydraulic Fluids

The working fluid for hydraulic installations is a nearly incompressible liquid and is required to have at the same time good lubrication properties. One uses today special hydraulic oils as supplied by the industry, whilst formerly engine oil was employed. The lubricating properties depend essentially on the fact that the oil adheres well to the walls or, in other words, is well wetting the parts with which it is in contact.

The most important characteristic variable of an oil is the *dynamic viscosity,* the effects of which will be described more in detail in chapter 2. It is a property of the material decreasing rapidly with increasing temperature. On the other hand, the viscosity increases with large static pressures, but this effect becomes noticeable only with pressures of several hectobars (several thousands of psi).

The viscosity of a given oil is, therefore, a function of the two variables temperature and pressure. There exist several more or less complicated mathematical expressions for this, but for hydrostatic machines the following formula is quite adequate

$$\mu = \mu_0 \, e^{-\frac{T}{T_x}} \, e^{\frac{P}{P_x}} \tag{12-1}$$

We see from formula 1 that the viscosity increases exponentially with the pressure and decreases exponentially with the temperature. This can be deduced theoretically from the Eyring theory of liquid viscosity. [1]

The symbols $T_x$ and $P_x$ in equation 1 denote a characteristic pressure and a characteristic temperature which one determines best from the measured viscosity versus pressure and temperature curves of a given oil. Naturally, an expression such as equation 1 has only a limited range of validity which, in this case, extends from $0°$ to about $100°$ C or well below the boiling point of the oil. The pressure range in which equation 1 is accurate goes from about 1 bar (15 psi) to about 1 000 bars (15 000 psi).

[1] — See for instance D. Tabor, 'Gases, Liquids and Solids', Penguin Library, 1969.

Further important properties of hydraulic oil are its compressibility and its thermal expansion. These effects are expressed by the *equation of state*. There are many such different equations, more or less complicated, but we shall use here a very simple one which is entirely satisfactory for the design and application of hydrostatic machines.

In order to develop an equation of state, we consider in fig 1 a vessel with thick walls and a hollow volume V filled with the oil to be examined. In order to apply a pressure and to measure the reduction in volume, one can use a tightly fitted piston as indicated on fig 1. It is loaded with a suitable force in order to generate pressure within the vessel and its movement indicates the reduction of volume. In order to exclude any perturbation by friction forces, it is possible to have the piston rotate slowly in its packing.

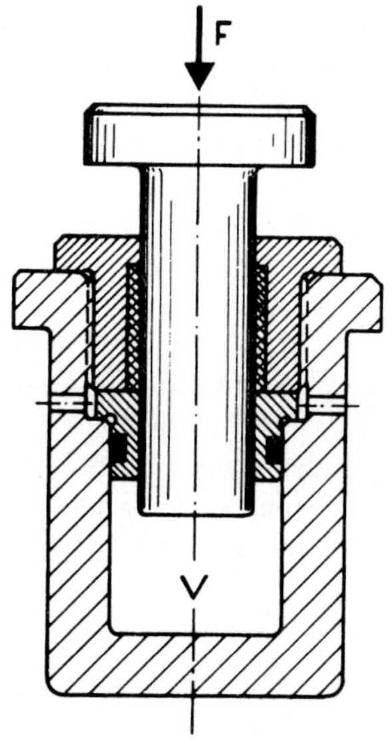


Fig 12.1: Pressure chamber for measuring compressibility

n the experiment of fig 1, one finds that in the pressure range between 0 and 500 bars (0 and 7 000 psi) the volume reduces approximately linearly with the applied pressure if temperature is maintained constant.

This can be represented mathematically as follows with the volume reduction $\Delta V$ and the pressure p

$$\frac{\Delta V}{V} = - \frac{p}{B} \qquad\qquad (12\text{--}2)$$

Equation 2 can be considered as the definition of the compressibility modulus B which has the dimension of a pressure. FN².

The compressibility modulus is strongly reduced by small amounts of air carried by the oil, for instance in the form of bubbles, but an acceptable approximate value for technically well deaerated oil would be

B = 150 hebar (= 220 000 psi)

This value of the compressibility modulus signifies that the volume reduces by only 1% when applying a pressure of 150 bar (2 200 psi). In the entire range of validity of equation 2 between 0 and 500 bar (7 000 psi), the volume reduction is only about 3.3%. Going to higher pressures, the oil becomes stiffer, ie it is reducing less in volume than would be indicated by equation 2.

It should be particularly noted that the slight volume reduction of 3.3% at 500 bar (7 000 psi) means that the oil is almost incompressible. This is quite different with gases which have, according to the so-called ideal gas equation, not only a much larger compressibility, but they also become much stiffer with increasing pressure.

By increasing the temperature in the arrangement of fig 1, one obtains in the technically important range from 0 to 100° C a linear expansion or volume increase. Therefore, equation 2 can be completed as follows

$$\frac{\Delta V}{V} = - \frac{p}{B} + \frac{\Delta T}{T_{xp}} \qquad\qquad (12\text{--}3)$$

Here, T is the temperature increase and $T_{xp}$ a constant of the dimension of the temperature which describes the thermal expansion, a typical value of which would be $T_{xp}$ = 1 500° C.

---

FN² — Therefore, it would be more appropriate to denote it by the letter p together with a suitable index, but B is internationally accepted.

Applying equation 3, it is to be noted that it is valid only until about $100^\circ$ C because, otherwise, one would come too near to the boiling point of the oil. For example, with a temperature increase of $75^\circ$ C

$$\frac{\Delta T}{T_{xp}} = \frac{75}{1500} = 0.05 \tag{12-4}$$

one obtains about 5% increase in volume.

Sometimes, it is convenient to use the mass density instead of the volume and we shall transform equation 3 accordingly. If the volume of the cavity in the vessel is constant, the relative reduction of mass density is equal to the relative increase of the volume

$$\frac{\Delta \rho}{\rho} = - \frac{\Delta V}{V} = \frac{p}{B} - \frac{\Delta T}{T_{xp}} \tag{12-5}$$

For very accurate measurements, it is necessary to consider the elastic deformation and the thermal expansion of the vessel which results in a correction of only about 5% due to elasticity and of 6% due to thermal expansion. In order to determine accurately the elasticity of a given vessel, one can calibrate it beforehand by making the compressibility experiment with mercury. As it is known from physics, the compressibility modulus of mercury is very constant and has the value of 2650 hebar (= 265 000 bar).

Fig 2 represents the viscosity as function of the temperature for different commonly used oils. FN[3]. We see from fig 2 how much viscosity decreases with increasing temperature. This effect severely limits the accuracy of all calculations where constant temperature and viscosity are assumed.


## Cavitation

Our equation 2 was set up for an applied or external pressure and all our calculations use the ambient air pressure as zero or reference point of the hydraulic pressure. Equation 2 is valid also for absolute pressures below the air pressure as long as the absolute pressure in the liquid does not fall below its vapour pressure. The absolute pressure is here the sum of the air pressure and of the external pressure.

If the absolute pressure decreases to near or below the vapour pressure, the liquid develops cavities or bubbles filled with vapour.

FN[3] — From information supplied by Mobil Oil (Switzerland)

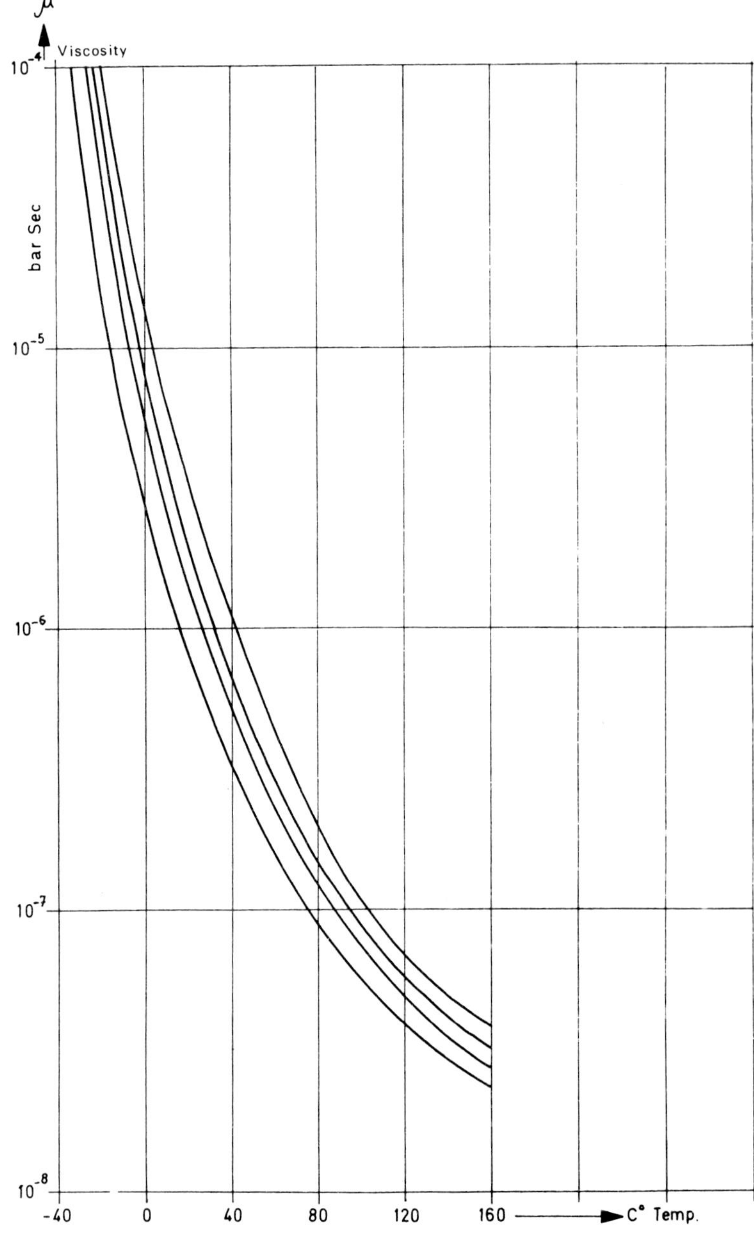


Fig 12.2:  Viscosity of usual hydraulic oils as function
of  temperature

This is called cavitation. Due to the surface tension of the cavity, especially important with very small bubbles, the onset of cavitation is somewhat irregular. Also dissolved gases may come out from the liquid at low absolute pressures, further influencing the beginning of cavitation.FN[4]

The vapour pressure of hydraulic oil is very small but increases rapidly with temperature. Table 1 contains some values of the vapour pressure at different temperatures. FN[5]

### Table 12—1

Vapour pressure of a typical hydraulic oil as function of temperature

| $T =$ | 20 | 40 | 60 | 80 | 100 | 120 | 140 | $^\circ$C |
|---|---|---|---|---|---|---|---|---|
| $p =$ | $9.10^{-5}$ | $6.10^{-4}$ | $4.10^{-3}$ | $2.10^{-2}$ | $8.10^{-2}$ | $3.10^{-1}$ | $8.10^{-1}$ | m bar |

Cavitation exists not only in hydrostatic machines, but also in water turbines and ship propellers. It is a damaging phenomenon and must be avoided. Cavitation produces loud  noises and shocks when the cavities collapse again under applied pressure. According to newer investigations, the fluid particles in the walls attain very high speeds which generate local high overpressures on impact on the solid walls eroding the material.

In oil hydraulic engineering, pressures below atmospheric pressure are generated usually if a pump has to draw oil from a tank with a free surface. In order to avoid cavitation, all pressure losses, from the intake tube to the pump cylinders, must be less than the air pressure. Otherwise, there would be cavitation beginning in the cylinders of the pump with its objectionable effects. It is, therefore, recommended to make the intake tube as short and as wide as possible and to keep exact account of the different pressure losses in the entire intake.

If intake pipes are below atmospheric pressure, they can sometimes produce vibrations and perturbations similar to cavitation. This

FN[4] — According to indications of Professor T. ishihara, Tokyo, small air bubbles are appearing in the hydraulic oil Teresso 52 with an absolute pressure of 300—400 mbar absolute (4.5—6 psi) at a temperature of $40^\circ$—$60^\circ$ C. If the absolute pressure is reduced to 30—50 mbar (0.45—0.7 psi), the oil boils violently. Refer also to Professor Ishihara's book, 'Oil Hydraulics', Tokyo, 1968, fig 4.69, page 125.

FN[5] — From information supplied by Mobil Oil (Switzerland).

is the case if air penetrates through minute openings or leaks. With
the beginning of the delivery stroke of the hydrostatic machine, these
air bubbles collapse unless they can be dissolved directly in the oil
and have the same damaging effects as cavitation. Apart from very
careful sealing of the intake tube, an old trick to avoid these troubles
is to arrange all pipes with less than atmospheric pressure in a tank
below the oil level.

The Thoma cavitation number /FN[6]/ describes the ratio of the difference
between the minimum static pressure and the vapour pressure to the
dynamic pressure in a component. It is very useful to compare experiences
between different models of water turbines (and other hydrodynamic machines). On the other hand, the Thoma cavitation number is less useful in oil
hydraulic engineering due to the great influence of viscosity.

## Displacement and Compression Energy

A flow under pressure transports energy in two forms that should
be clearly distinguished. These forms are:

1.  The displacement energy needed to displace the fluid against
    the opposing hydrostatic pressure. It is the principal energy
    carrier in oil hydraulics.

2.  The compression energy contained in the compressibility of the
    fluid, as in a loaded spring. It is almost always negligible in
    oil hydraulics FN[7], but important in pneumatics and heat
    engines.

Fig 3 illustrates the effect of the two energies. If we admit into
a suitable chamber, such as a cylinder with piston, volume $V_o$ under
pressure $p_o$, the displacement energy $E_{dis} = p_o V_o$ becomes available, for instance, at the piston. Closing the fluid admission, the
oil expands under decreasing pressure, as shown by the inclined
part of fig 3, giving up the compression energy. Due to the linear
pressure drop according to equation 2, the compression energy
corresponds to the triangular area in fig 3 and is given by

$$E_{comp} = \tfrac{1}{2} \Delta V\, p_o = \tfrac{1}{2} V\, \frac{p_o^2}{B} \qquad (12\text{--}6)$$

FN[6] — After D. Thoma, the late uncle of the author.

FN[7] — Exceptions are the hydraulic presses in section 53 and the shock waves
in section 73.

The ratio between compression and displacement energy becomes

$$E_{comp} / E_{dis} = \tfrac{1}{2} \; \frac{p_o}{B} \tag{12-7}$$

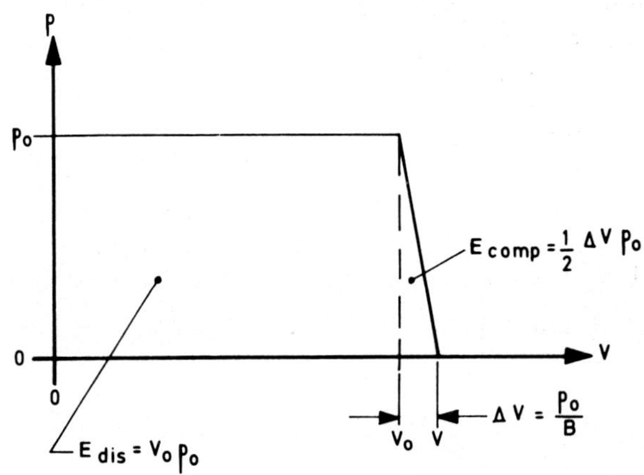


**Fig 12.3: Diagram with displacement and compression energy**

With a usual compressibility modulus of $B = 150$ hebar, this ratio is very small, of the order of 1% with a pressure of 300 bar. This is the reason why it is negligible, but we must not forget its existence in principle. Furthermore, the difference of volume flow at different pressures is normally neglected in oilhydraulics but must be considered when making accurate statements about losses and efficiency.

## 13    Flow under Inertia Forces

In hydrostatic machines and installations, the forces due to the pressure gradient (due to the change of pressure over a given length) and due to the viscosity are usually preponderant. Nevertheless, we shall treat here briefly the effect of inertia forces, ie the forces depending on the acceleration of the flow.

Let us consider in fig 1 a certain tube of length L and cross section A which is supplied from the left from a large tank so that an average flow speed $v_m$ is obtained. We would like to determine the power needed to accelerate the liquid in the tube. The kinetic energy of the oil volume is

$$E = m \; \frac{v_m^2}{2} = \rho V \; \frac{v_m^2}{2}$$

and the necessary power for the acceleration of the continuously supplied volume

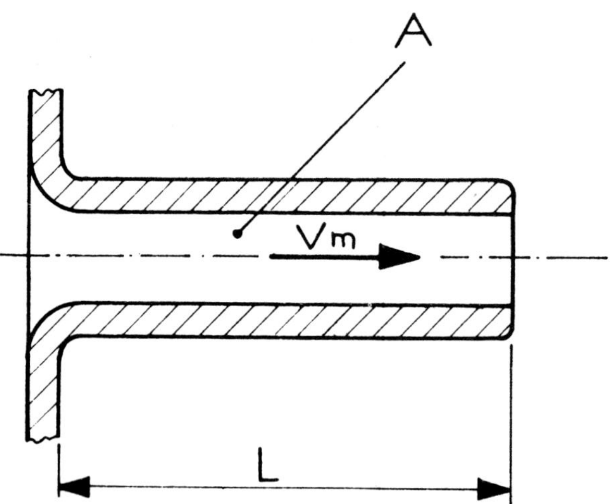


**Fig 13.1: Tube for the calculation of the power needed for acceleration**

$$P = \frac{dE}{dt} = \rho \ \frac{dV}{dt} \ \frac{v_m^2}{2} = \rho Q \ \frac{v_m^2}{2} = P_{dyn} \, Q. \tag{13-1}$$

The power is proportional to the volume flow according to the last form of equation 1 multiplied with a variable of the dimension /FN[1]/ of a pressure. We shall call the latter the dynamic pressure, allowing us to write

$$P_{dyn} \qquad \rho \ \frac{v_m^2}{2} \quad \text{and} \quad P = P_{dyn} \, Q \tag{13-2}$$

With an additional static pressure in the tube, the entire power conveyed is /FN[2]/

$$P_{tot} = P_{stat} \, Q + P_{dyn} \, Q \tag{13-3}$$

The notion of the dynamic pressure allows us to define hydro-statics as the part of fluid mechanics in which the static pressure is much larger than the dynamic pressure in normal operation. Hydro-dynamics, on the other hand, are characterised by much higher dynamic than static pressures.

### Numerical Example of Dynamic Pressure

What is the dynamic pressure of a flow with $v_m$ = 10 m/sec = 1 000 cm/sec and a mass density of $\rho = 0.9 \cdot 10^{-3}$ kg/cm?

$$p = 0.9 \cdot 10^{-3} \ \frac{kg}{cm^3} \ \frac{10^6}{2} \ \frac{cm^2}{sec^2} = 450 \ \frac{kg}{cmsec^2} = 450 \text{ mbar}$$

For the last form of the above equation, we have used from the numerical example of section 11

$$1 \ \frac{kg}{cmsec^2} = 1 \text{ mbar}$$

The dynamic pressure of a flow velocity of 10 m/sec is less than 0.5 bar.

For other velocities, we can write the following form, valid for liquids of the given mass density

---

FN[1] — As explained in more detail in appendix 3, one would better use the word 'class' instead of 'dimension , but we shall adopt here the usual terminology.

FN[2] — Strictly the compression energy of the fluid has to be added, but it vanishes with incompressible fluid.

$$\frac{p}{0.45 \text{ bar}} = \left[\frac{v_m}{10 \text{ m/sec}}\right]^2 \qquad (13\text{--}2b)$$

Equation 2b can be expressed in words as follows

The dynamic pressure is as many times 450 mbar as the velocity a multiple of 10 m/sec, the latter taken squared'.

We should note that the liquid is again decelerated when leaving the tube, liberating the power that was needed for acceleration. As shown in general fluid mechanics, we can recover this power in principle by a suitable slow increase of the cross section of the tube re-establishing a corresponding static pressure. If this suceeds, the relation between static and dynamic pressure is given by the well known Bernoulli equation, but this is outside the scope of the present book. In hydrostatics, it is usually not worthwhile to provide the slowly increasing cross section for power recovery so that at least a part of it is lost.

Our model of fig 1 is incomplete in that the viscosity forces have been neglected. It would be valid in principle for a vanishing viscosity but it is known that the flow becomes unstable in this case. Instead of an orderly flow, also called laminar flow as treated in chapter 2, we obtain an irregular, turbulent flow with a strong mixing of the fluid particles. The limit of stability does not depend on viscosity alone but on the following dimensionless product expression, generally called Reynold's number

$$\text{Re} = \frac{\rho \, v_m \, L}{\mu} \qquad \text{Re} > 2\,300 \rightarrow \quad \text{means turbulence} \qquad (13\text{--}4)$$

Here L is a characteristic length, for instance the diameter of the tube. The value of Re = 2300 is valid only for straight and polished tubes but it is a useful approximation of hydrostatic components. In reality, the stability limit depends very much on design particulars, for instance the rounding of the inlets.


**Numerical Example of Reynold's Number**

What is the Reynold's number in a tube with a diameter of 1 cm and with $v = 100$ cm/sec, $\mu = 2 \cdot 10^{-7}$ barsec /FN³/ $\rho = 0.9 \cdot 10^{-3}$ kg/cm³?

FN³ — Barsec is the abbreviation of daNsec/cm² as explained in section 21.

$$Re = \frac{0.9 \cdot 10^{3} \text{ kg/cm}^{3} \cdot 100 \text{ cm/sec} \cdot 1 \text{ cm}}{2 \cdot 10^{-7} \text{ barsec}}$$

$$Re = 0.45 \cdot 10^{6} \; \frac{\text{kg cm}}{\text{sec}^{2} \text{daN}}$$

We know from the numerical example in section 11 that

$$\frac{1 \text{ kg}}{\text{cmsec}^{2}} = 10^{-3} \; \frac{\text{daN}}{\text{cm}^{2}}$$

and we obtain by inserting

$$Re = 0.45 \cdot 10^{6} \cdot 10^{-3} = 450$$

For different flows but with the given values of viscosity and mass density, we can set up the following practical formula. We compute first Reynold's number with $v = 1$ cm/sec   and $L = 1$ cm

$$Re = 0.45 \cdot 10^{6} \cdot 10^{-3} = 450$$

With higher velocities and diameters (characteristic length $\cdot$ L), the Reynold's number increases proportionally allowing us to write

$$Re = 450 \, \frac{v_{m}}{1 \text{ m/sec}} \; \frac{L}{1 \text{ cm}} \qquad\qquad (13\text{--}4b)$$

This can be expressed in words as follows.

With normal hydraulic oil, the Reynold's number is equal to 4.5 times the velocity, expressed in cm/sec, times the diameter in cm.

This formula is especially useful for quick order of magnitude calculations.

The pressure drop associated with flow through components like tubes, valves, elbows is usually given as a multiple of the dynamic pressure, with a loss coefficient $\zeta$, as follows/FN[4] /

---

FN[4] — In the author's opinion, this way of representing the pressure drop is more useful with turbulent than with laminar flow.

$$\Delta p = \zeta\, p_{dyn} = \zeta\, \frac{\rho}{2}\, v_m^2 = \zeta\, \frac{\rho}{2} \left[\frac{Q}{A}\right]^2 \qquad (13\text{--}5)$$

Q  =  oil flow
A  =  cross section

With turbulent flow, the loss coefficient $\zeta$ decreases only slowly with increasing Reynold's number as treated in the literature. /FN[5]/ For approximate calculations, we can use

$\zeta = 0.5 - 1$ in elbows

$\zeta = 1 \quad - 2$ in valves

It should also be noted that the loss coefficient can be substantially reduced by rounding or chamfering the edges of flow restrictions.

In pipes, the pressure losses are continuously distributed and, therefore, proportional to the length of the pipe. This can be represented with a friction factor $\lambda_R$ as follows

$$\Delta p = \lambda_R\, \frac{L}{d}\, \frac{\rho}{2} \left[\frac{Q}{A}\right]^2 \qquad (13\text{--}6)$$

where L = length and d = diameter of the tube.

The friction factor $\lambda_R$ depends on the Reynold's number and on the surface roughness as explained in the literature. In oil hydraulics with usually flat surfaces, we can use, with turbulent flow, an approximate value of $\lambda_R = 1/40$. Then, equation 5 can be put in words as follows:

'The pressure loss of turbulent flow in the pipe is equal to one dynamical pressure for each length equal to 40 diameters'.

With a round tube, we obtain with $A = \frac{\pi}{4}\, d^2$

$$\Delta p = \frac{16}{\pi^2}\, \frac{L}{40}\, \frac{Q^2}{d^5}\, \frac{\rho}{2}$$

FN[5] — See Guillon lit.7.

FN[6] — With laminar flow, one has $\lambda_R = 64/Re$ which implies a pressure loss proportional to oil flow.

   With a given turbulent oil flow, the pressure loss is inversely proportional to the fifth power of the diameter and increases very rapidly with decreasing diameter.

   It must not be forgotten, however, that with laminar flow the pressure loss is always the sum of the viscosity effects, proportional to flow velocity, and of the inertia or acceleration effects, the latter proportional to the square of the velocity. Therefore, the characteristic curves of pressure versus flow have always a linear part near the origin where the viscosity effects are preponderant. This is valid also with valves and elbow pieces even if one loses several dynamical pressures with normal or rated flow.

## Impulse Effects of Oil Jets

   We shall now study the force exerted by a jet impinging on a plate. For this, we use fig 2 with a well formed nozzle at the left fed with the pressure $p_\delta$ whilst the plate is on the right. /FN[7]/ The oil flow is connected with the cross section of the nozzle as follows

$$Q = A\, v_m$$

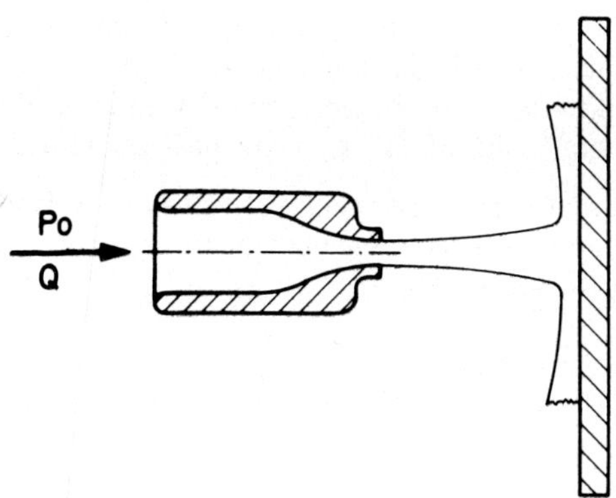


**Fig 13.2: Cross-section through nozzle, jet and plate
for the calculation of momentum**

---

FN[7] — With servovalves, the plate is called a flapper.

If the supply pressure exceeds several bars (several tens of psi), the acceleration forces in the nozzle are the most important and the mean velocity $v_m$ is related to the applied pressure $p_o$ as follows

$$p = \zeta \; \frac{\rho}{2} \; v_m^2 \qquad\qquad (13\text{--}7)$$

where $\zeta$ is the loss coefficient. It indicates not only the effect of wall friction but also that the jet may not entirely fill the cross section suffering from the so-called contraction. With well rounded nozzles, these effects are small and we can approximately set $\zeta = 1$.

The force of the impinging jet is equal to the change of its momentum

$$F = \frac{d}{dt} \; m \; v_m \;=\; \rho \; \frac{dV}{dt} \; v_m \;=\; \rho \; A \; v_m^2 \qquad\qquad (13\text{--}8)$$

We obtain by inserting equation 7

$$F = 2 \, A \, p_o \qquad\qquad (13\text{--}9)$$

We have now obtained the surprising result that the force (with $\zeta = 1$) is *twice* as large as the product of supply pressure and cross section of the nozzle. On the other hand, if we approach the plate to the nozzle until there is no longer much oil flow, the plate experiences only the hydrostatic force $F = p_o A$, ie half as large. This reduction can lead to instabilities with movable plates or flappers in servo-valves. For a more detailed treatment of forces, the reader is referred to Bibliography 1, section 10.4.

## 14   Hydraulic Circuit Techniques

### 14–1   Hydraulic Dipoles

In hydraulic installations, many components are connected by tubes or pipes, thus building entire circuits as in electronics. We shall treat here the principles of hydraulic circuit techniques.

The fundamental hydraulic component is the dipole, simply an arbitrary component with two hydraulic connections. The operation of a dipole is described by its characteristic curves or characteristics of the relation between volume flow and pressure drop, independently of constructional details. Examples of hydraulic dipoles are : Throttles, relief and flow control valves, but also pieces of piping and all other components with a definite relation between flow and pressure drop.

Hydrostatic machines have also two main connections, but they can be considered as dipoles only within limits since the hydraulic characteristics depend on the mechanical drive behaviour, as we shall see in section 62.

Hydraulic power units supplying a constant pressure almost independent of flow or a constant flow almost independent of pressure can be treated as idealised dipoles.We shall refer to them as pressure sources or flow sources respectively.

For describing the circuits of hydraulic installations, there are a number of internationally agreed circuit symbols as summarised in appendix 1.

### 14–2   Series and Parallel Connections

All circuits are built by a combination of series and parallel connections of hydraulic components. The principle of series connection is represented in fig 1 where both rectangles can be any

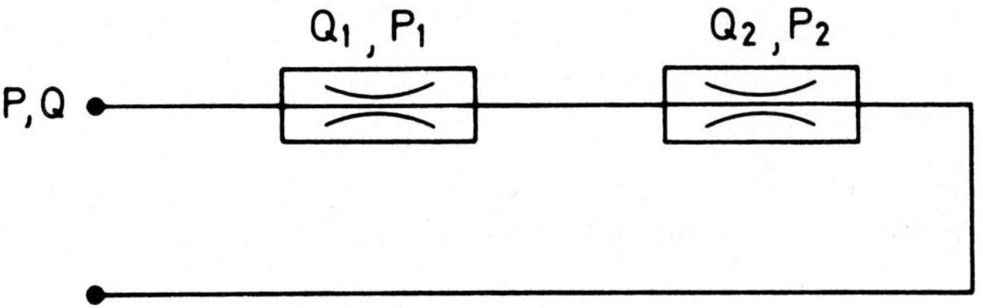


**Fig 14.1: Series circuits of dipoles**

dipoles. Let us suppose for definiteness that we have two throttles as indicated by the rectangle as circuit symbol. The important property of the series connection is that the flows are equal and the pressure drops of the dipoles are additive as follows

$$p = p_1 + p_2 \qquad Q = Q_1 = Q_2 \qquad (14-1)$$

The parallel connection, on the other hand, is represented in fig 2. Here, the pressure drops on both dipoles are equal but the volume flows are additive

$$p = p_1 = p_2 \qquad Q = Q_1 + Q_2 \qquad (14-2)$$

It should be noted that these relations are also valid if the dipoles have non-linear characteristics.

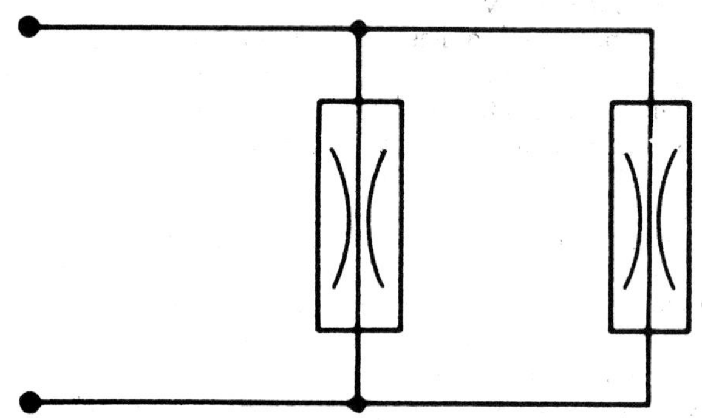

**Fig 14.2: Parallel circuit of dipoles**

### 14-3   Hydraulic Impedances and Admittances

The characteristics of dipoles determine the relation between flow and pressure drop. We can thus select at will one of these variables, but then the other one is determined from the characteristics. Consequently, we have two different ways of describing the operation of dipoles:

1. We impose the flow and read the pressure drop from the characteristics.
2. We impose the pressure drop and determine the flow from the characteristics.

Both choices are represented in fig 3 as functional diagrams with the appropriate input and output variables. /FN[1] /.

FN[1] — Functional diagrams are a powerful tool with hydrostatics and other branches of engineering. Their use is explained in appendix 2.

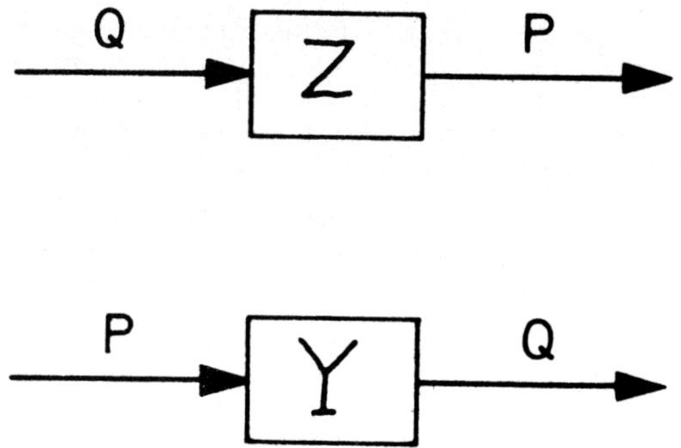



**Fig 14.3: Elements of functional diagram with two
different causalities**

In graphical representations or graphs, it is usual to represent
the independent variable on the horizontal or x axis and the dependent
variable on the vertical or y axis. Adopting this usage, we can either:

1.  Represent the imposed flow on the horizontal axis and the
    consequent pressure drop on the vertical axis and call this the
    Z representation.
2.  Represent the imposed pressure drop on the horizontal axis
    and the consequent flow on the vertical axis and call this the
    Y representation.

These two possible choices of cause and effect are also called
*selection of causality*. The change from one to the other causality is
called inversion, a very useful concept with functional diagrams.

As explained in detail in appendix 2, the inversion of functional
diagrams is practically always possible. For the study of hydrostatic
machines, it is advantageous sometimes to use the one, sometimes
the other causality. In other words, we shall consider as given sometimes the flow and sometimes the pressure of a dipole and calculate
the other variable from the characteristics.

The variables flow and pressure are also called complementary variables
since their product is a power. The same is true for rotation frequency and
torque, speed or velocity and force, electric current and voltage.

This relation is used to set up a further graphical representation, bond graphs, after Professor Paynter. /FN$^2$/ They allow establishment of functional diagrams with different causalities by study of the power flows. Bond graphs are especially useful for the study of complex installations but we shall not need them in this introductory course.

In order to express the characteristics of a dipole by a physical variable, we shall introduce impedances and admittances, as known for linear components, as follows

$$p = Z \cdot Q \quad \text{or} \quad Z = p/Q = 1/Y \qquad\qquad (14-3)$$

$$Q = Y \cdot p \quad \text{or} \quad Y = Q/p = 1/Z$$

The dimension of a hydraulic impedance is

$$\dim Z = \dim p/Q \;=\; \frac{F/L^2}{L^3/t} \;=\; \frac{Ft}{L^5} \;\to\; \frac{bar}{cm^3/sec}$$

In order to use the concept of impedances in hydrostatics, where components usually have non-linear characteristics, we consider fig 4.

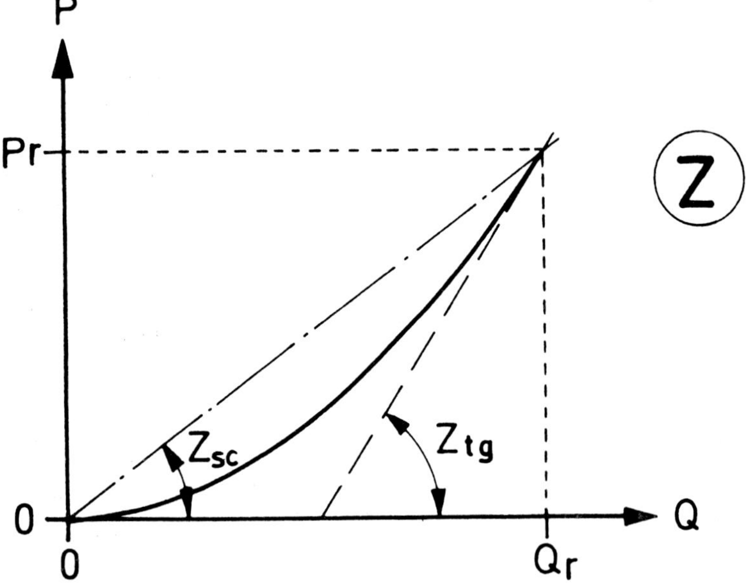


Fig 14.4:  Z  representation of the  characteristics  of
a dipole

FN$^2$ — See for instance D. Karnopp and R. Rosenberg, 'Analysis of Multiport Systems', MIT Press, 1968.

It shows the curved characteristics of a throttle in Z representation where the pressure drop is an increasing function of the oil flow and also an operating point $p_r/Q_r$.

For small deviations from the operating point, we can replace the characteristics by its tangent and represent the pressure increase as follows

$$\Delta p = Z_{tg} \, \Delta Q \quad \text{or} \quad Z_{tg} = \frac{\Delta p}{\Delta Q} \tag{14–4}$$

Equation 4 is the definition of $Z_{tg}$, the *tangent impedance*. Its name is derived from the fact that it corresponds to the slope of the tangent.

On the other hand, it is desirable to introduce another variable, the *secant impedance*, for the relation of the variables in the operating point as follows

$$p_r = Z_{sc} \, Q_r \quad \text{or} \quad Z_{sc} = p_r/Q_r \tag{14–5}$$

The name secant impedance follows from the fact that it represents the slope of the secant between the operating point and the origin, as shown in fig 4 as point-broken line.

With the Y representation, it is convenient to introduce *tangent and secant admittances* as follows

$$\Delta Q = Y_{tg} \, \Delta p \quad \text{or} \quad Y_{tg} = \Delta Q/\Delta p = 1/Z_{tg}$$

$$\tag{14–6}$$

$$Q_r = Y_{sc} \, p_r \quad \text{or} \quad Y_{sc} = Q_r/p_r = 1/Z_{sc}$$

Also in Y representation, the tangent and secant impedances represent the slope of the tangent and of the secant of the curve between the origin and the actual operating point, as we can see from fig 5.

It should be noted that both impedances and admittances depend with non-linear characteristics on the operating point, as can be seen easily by drawing tangents and secants in fig 4 to other operating points.

Components with linear characteristics going through the origin or zero point — as for instance with the oil filled gap as treated in section 22 or with ordinary (ohmic) resistors in electronics — have equal tangent and secant impedances. This comes from the fact that a straight characteristic through the origin coincides both with its tangent and secant.

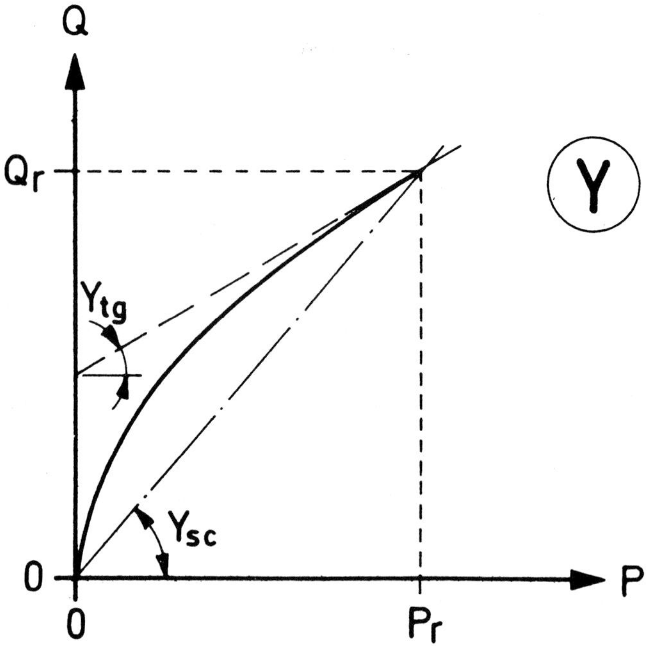


**Fig 14.5:** Y representation of the characteristics of
a dipole

With a linear characteristic not going through the origin, the tangent impedance remains constant but the secant impedance depends on the operating point.As an example, in fig 6 is shown the characteristics of a pressure source with a certain linear internal resistance. This shows clearly the actual characteristics of hydraulic power packs where the nominally constant pressure decreases slightly with large oil flow.

Fig. 6. contains two operating points, 1 and 2, with their secants whilst the tangents coincide with the linear characteristics. We see that with smaller flow, point 2, the secant becomes steeper or the secant impedance increases. We would find exactly the same conditions in electrotechnics if one describes the characteristics of a DC generator or of a battery with internal resistance.

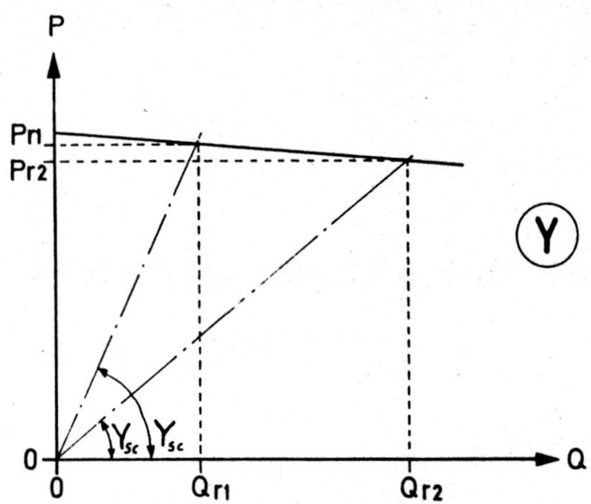


**Fig 14.6:** Z representation of the characteristics of a hydraulic power pack with nominally constant pressure

The distinction of tangent and secant impedances is a very useful method, not only in hydrostatics but also in other parts of engineering with non-linear characteristics, and even in economics. In the latter case, instead of the pressure one uses the price of a commodity and instead of the flow the total value of the production of said commodity.

We shall use tangent and secant impedances frequently, especially when describing the operation of relief valves and flow control valves in sections 42 and 43.

# 2   THE GAP AS BASIC HYDROSTATIC ELEMENT

### 21   Flow in Gaps

Many parts of hydrostatic machines and components are separated by oil-filled gaps essential for trouble free performance. In this context, a gap is a space limited by metallic walls that is in one direction, the gap height, much smaller than in the other directions called length and width. Typical values for the gap height in oil hydraulics are 5 — 20 micrometers /FN[1]/, whilst the length and width of the gap are in the order of 5 millimeters (0.25 inches) and more. Otherwise, the gap can have any form; it does not have to be flat, but it may be curved or rolled up, as long as the rolling radius is small compared to gap height.

We find gaps in oil hydraulics principally between pistons and cylinders, but also between the rotor and the casing of vane pumps (see section 34). Further examples are axial and radial plain bearings under full fluid lubrication. There, hydrostatic pressure in the gap transmits the bearing forces between the shaft and the fixed part. Even in ball and roller bearings there is some kind of gap with a pressure field near to the contact point, as we shall see in section 27.

Gaps are thus a basic element influencing the design of many hydrostatic machines. We shall, therefore, discuss the different forms of gaps in this chapter including, in section 25 and 26 briefly, the hydrodynamic and hydrostatic bearing design, useful also outside oil hydraulics.

Generally, flow and pressure development in a gap are influenced by the following factors:

---

FN[1] — 1 micrometer $= 10^{-3}$ millimeter $= 10^{-6}$ meter $= 40$ microinches.

1. Wall movement

   The walls of the gap can be stationary or they can have relative movement.

2. Gap height

   The gap height can be constant or it may be variable due to nonparallel walls.

3. Pressure drop

   At the beginning and at the end of the gap, there can be the same pressure or there may be a higher pressure, for instance at the beginning.

A piston in a cylinder of an axial piston machine gives us examples for all these cases. If the piston is stationary, the gap walls do not move but, normally, their relative velocity is equal to the piston speed. The gap height is constant only if the piston is exactly cylindrical whilst a slightly conical piston (or cylinder) leads to strongly variable gap height. The same effect exists if a cylindrical piston is in a slightly oblique position in the cylinder. The applied pressure is equal to the operating pressure of the axial piston machine. It vanishes if the flow requires no pressure, which happens from time to time during a typical operating cycle.

All important cases of gap flow can be composed from the above possibilities, as we shall describe below. We shall take the oil as incompressible and examine the relations between the pressure drop and the flow. Due to the assumed incompressibility, the volume flow remains constant within the gap.

In mathematically more rigorous form, one describes the gap flow by integration of the Navier-Stokes equations of fluid mechanics introducing the following approximations:

1. The inertia forces are neglected or the mass density is set $\rho = 0$.

2. The fluid velocity in direction of the gap height is zero.

3. The gap is infinitely large and the velocity independent of the gap width.

4. The liquid is incompressible.

With these assumptions, the Navier-Stokes equations reduce to

$$\frac{dp}{dz} = -\frac{d}{dx}\mu\frac{dv}{dx}$$

This is still a partial differential equation comprising derivations with respect to x, the co-ordinate in the direction of the gap height, and with respect to z, the co-ordinate along the gap. The boundary conditions are derived from the fact that the oil adheres to the walls giving a condition for the flow velocity depending on the wall movement. The volume flow, furthermore, is constant which means

$$Q = b \int_{0}^{h} v \, dx = \text{constant}$$

It is interesting to note that the above equations are also valid with variable viscosity, for instance, due to heating of the oil along the gap.

## Gap with Parallel Walls without External Pressure

A simple form of gap flow has parallel, moving walls without external pressure. We shall use it here, as in most books on elementary physics, to introduce the notion of viscosity.

Consider in fig 1 a gap with the height h where the lower wall runs with the velocity $v_0$ to the right. The upper wall is in proximity of the lower wall within a portion of the length L, thereby defining the gap length. We use a system of co-ordinate with the x co-ordinate in the direction of the height and the z co-ordinate in the direction of the length, i.e. of the movement of the lower wall.

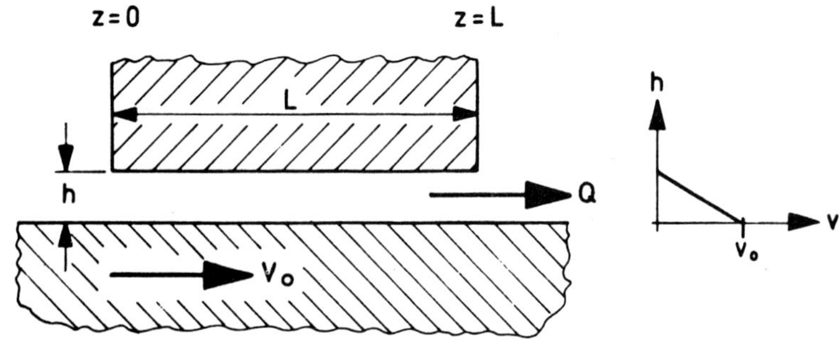


Fig 21.1: Cross-section of a gap with moving lower
wall (left) and velocity profile (right)

Since the gap height is very small (h = 5 — 20 micrometers or 200 — 800 microinches), the flow is essentially controlled by the viscosity of the liquid as follows:

1. The liquid adheres to the walls and assumes the wall velocity in their immediate vicinity.

2. There is a shear tension between the fluid layers proportional to the gradient of velocity, i.e. the derivative of the velocity with respect to the x co-ordinate.

We can therefore represent the shear tension $\tau$ as follows

$$\tau = -\mu \frac{dv}{dx} \qquad (21-1)$$

Here, $\mu$ denotes the viscosity which has the following dimension

$$\dim \mu = \frac{F/L^2}{(L/t)L} = \frac{Ft}{L^2}$$

Since we use in our calculations as unit of length the centimetre, as unit of time the second and as unit of force the decanewton, we obtain as unit for the viscosity the expression daNsec/cm². Since 1 daN/cm² is equal to 1 bar /FN²/, we can call the unit of viscosity also barsec which has the advantage of easy pronounciation. It can be considered as an abbreviation for the above product expression and it indicates how the viscosity can be divided out against a pressure when making dimensional considerations. Therefore, we have

$$\text{Unit of } \mu : \quad 1 \frac{\text{daNsec}}{\text{cm}^2} = 1 \text{ barsec}$$

In the literature, one also uses the following unit of viscosity

$$1 \text{ poise} = 10^{-4} \frac{\text{daNsec}}{\text{cm}^2} = 1 \text{ microbarsec} = 1 \ \mu\text{barsec}$$

and also

$$1 \text{ centipoise} = 10^{-8} \frac{\text{daNsec}}{\text{cm}^2} = 10^{-8} \text{ barsec}$$

For approximate calculations, we shall use 0.2 $\mu$barsec as representative value.

---

FN² — We also use here the bar for shear tension which is admissible as explained in appendix 3.

In general fluid mechanics, one frequently refers to kinematic viscosity which is the ratio between our viscosity $\mu$ and the mass density $\rho$, especially for determination of the Reynold's number. Our viscosity is then called dynamic viscosity (which the author considers as rather unfortunate terminology).

The kinematic viscosity has the dimension

$$\operatorname{dim} \frac{\mu}{\rho} = \frac{F\,t/L^2}{m/L^3} = \frac{F\,t\,L}{m} = \frac{L^2}{t}$$

Therefore, one obtains for the unit of the kinematic viscosity

$$1\ \frac{cm^2}{sec} = 1\ stokes = 100\ centistokes$$

*Numerical example:*  What is the kinematic viscosity of an oil with a (dynamic) viscosity of $10^{-6}$ barsec $= 1$ microbarsec $= 1$ poise and a mass density of $\rho = 0.9 \cdot 10^{-3}$ kg/cm$^3$ ?

$$\frac{\mu}{\rho} = \frac{10^{-6}\ daNsec/cm^2}{0.9 \cdot 10^{-3}\ kg/cm^3} = 1.1 \cdot 10^{-3}\ \frac{daNsec\ cm}{kg}$$

Since we have 1 daN $= 10$ kgm/sec$^2$ $= 1\,000$ kgcm/sec$^2$, we obtain

$$\frac{\mu}{\rho} = 1.1\ cm^2/sec$$

The viscosity of an oil with the given mass density in microbarsec (equal to poises) is about 10% larger than the kinematic viscosity expressed in stokes. It is useful to remember this relation since many tables and catalogues of firms contain only the kinematic viscosity in stokes.

If there is no external pressure applied to a gap — as we assume in this section — then the shear tension is everywhere equal and one obtains a constant velocity gradient, i.e. a linear increase of velocity across the gap height. This is represented in fig 1 at the right. Consequently, the velocity gradient is connected with the wall velocity $v_o$ and the gap height as follows

$$\frac{dv}{dx} = \frac{v_o}{h} \qquad\qquad (21-2)$$

We note in particular that, due to the shear tension and to the adhering of the liquid, the lower wall is braked and the upper wall is entrained or dragged along, a fact often overlooked in the literature. The force is equal to the product of the shear tension and the area of the gap wall, i.e. the length L and the width b, the latter perpendicularly to the plane of the paper in fig 1.

$$F = \tau\,L\,b \quad = \quad \mu\frac{v_o}{h}\,Lb \qquad\qquad (21-3)$$

Due to the linear increase, the average velocity is equal to one half of the wall velocity $v_o$. Therefore, the following volume flow is delivered to the right in fig 1.

$$Q \quad = \quad \frac{h\,b}{2}\,v_o \qquad\qquad (21-4)$$

It is important to note that the flow develops according to equation 4 only if there is no pressure at the end of the gap, which means that the oil can freely flow away. With a suitable restriction, on the other hand, high pressures up to more than 100 bar (1 500 psi) can build up. The gap, therefore, has a kind of pumping effect but with considerable energy losses. Therefore, designs based thereon are not used in practice.

Let us finally note that, due to the braking of the lower wall in fig 1, some power is destroyed or transformed into heat. It is given as the product of braking force and wall velocity

$$P \quad = \quad F\,v_o \quad = \quad L\,b\,\mu\,\frac{v_o^2}{h} \qquad\qquad (21-5)$$

The power is proportional to the square of the wall velocity and appears again as temperature increase of the wall and of the outgoing oil.

*Numerical example:*      A piston with 20 mm diameter is moving centrically in a cylinder with a diametral clearance of 20 micrometers = $2 \cdot 10^{-3}$ cm and with $v_o = 500$ cm/sec. What is the braking force and the delivery or flow of the gap ? Viscosity $\mu = 0.2$ microbarseconds = $2 \cdot 10^{-6}$ barsec.

The gap is rolled up, its width equal to the circumference of the piston.

$$b = \pi \cdot 2\ \text{cm} = 6.28\ \text{cm}$$

$$F = 2 \cdot 10^{-7} \, \frac{daNsec}{cm^2} \cdot 500 \, \frac{cm}{sec} \cdot \frac{10 \, cm \cdot 6.28 \, cm}{2 \cdot 10^{-3} \, cm}$$

$$F = 3.14 \, daN$$

$$Q = \frac{h \, b}{2} \, v_o = \frac{2 \cdot 10^{-3} \, cm \cdot 6.28 \, cm}{2} \, 500 \, \frac{cm}{sec}$$

$$Q = 3.14 \, cm^3/sec$$

It is now interesting to note that this braking force corresponds to a pressure of 1 bar with the piston area of 3.14 cm². If the piston is driven with a given force, the pressure produced is smaller by one bar than its theoretical value.

We note for comparison that the flow delivered by the piston area is as follows

$$Q = A_{\hat{p}} \, v_{\hat{o}}$$

$$Q_o = 3.14 \, cm^2 \, 500 \, \frac{cm}{sec} = 1570 \, \frac{cm^3}{sec}$$

The gap delivery is 3.14/1570 =.2% of the piston delivery. The gap delivery follows the piston movement and is not a leakage. It can, therefore, be taken into account by assuming an effective piston diameter comprising one half of the diametral clearance. Any leakage due to applied pressure will be treated in section 22.

## 22    Gaps with Parallel Fixed Walls

We shall consider now a gap with parallel and stationary walls through which the fluid is driven by an externally applied pressure. The oil adheres to the walls due to viscosity, strongly braking the flow. Thus, the average velocity and the flow remain small even with very high pressure gradients.

It is possible to use such gaps as sealing devices although, with a given applied pressure, a certain leakage flow always remains. Similar devices are found in steam turbines. However, there the mechanism of braking of the leakage flow is entirely different, depending on inertia effects, i.e. on the pressure required for acceleration of the fluid.

We shall now derive the so-called *gap formula* for the flow which is of fundamental importance for the design of hydrostatic machines and components.

We consider in fig 1 the longitudinal and cross section through a gap with height h and width b. The width is supposed to be much larger than the height, at least 10 times as large, or much more than represented on the figure. Therefore, the influence of the lateral walls can be neglected.

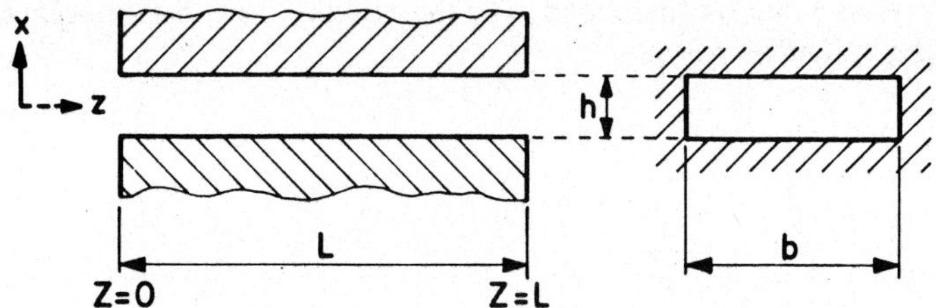


Fig 22.1:    Longitudinal and cross-section of a gap

The applied pressure $p_0$ pushes the oil into the gap from the left. Since the oil adheres to the upper and lower walls of the gap, there is a braking force transmitted between the fluid layers by the shear forces due to the viscosity. Considering a layer of fluid of the height 2x and width b, the equilibrium between the pressure force on the frontal area and the shear forces on the upper and the lower area can be written as follows

$$F = p_0 \, b \, 2x = 2\tau L b = -2\mu \frac{dv}{dx} L b \qquad (22\text{--}1)$$

$v = $ velocity along gap

With given applied pressure $p_0$, the shear force determines the derivative of the velocity with respect to the x axis. We obtain by equalising the second and the fourth member of equation 1

$$\frac{dv}{dx} = -\frac{p_0}{L}\frac{x}{\mu} \qquad (22\text{--}2)$$

The velocity distribution is obtained by integration over x

$$v = -\frac{p_0}{L}\frac{x^2}{2\mu} + \text{const}$$

The constant of integration is obtained from the condition that for $x = h/2$ the liquid adheres to the walls, implying that the velocity vanishes at that point

$$v = \frac{p_0}{2\mu L}\left[\left(\frac{h}{2}\right)^2 - x^2\right] \qquad (22\text{--}3)$$

Equation 3 describes the parabolic distribution of the velocity as known from physics textbooks. For the volume flow, we integrate over the gap height as follows

$$Q = b \int_{-\frac{h}{2}}^{+\frac{h}{2}} v \, dx$$

$$Q = \frac{1}{12\mu}\frac{p_0}{L} b \, h^3 \qquad (22\text{--}4)$$

Equation 4 is the very important *gap formula* which we shall encounter in many situations and, in particular, use it to calculate the leakage in hydrostatic components and machines. It gives a leakage flow proportional to the applied pressure and inversely proportional to the viscosity of the oil.

It is possible to use the gap formula in two different ways as represented in fig 2 as functional diagrams, /FN[1]/.

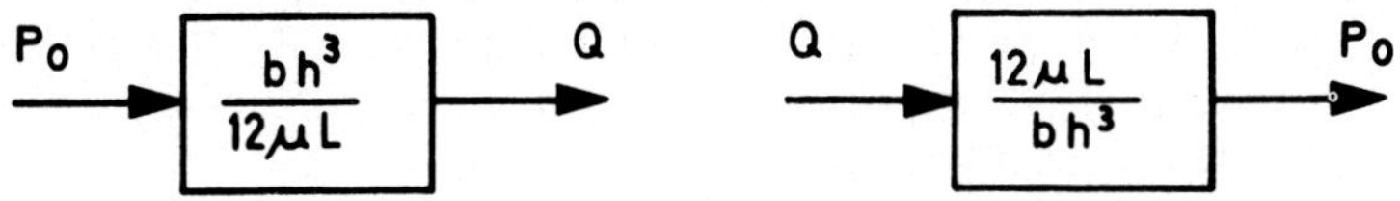


Fig 22.2: Functional diagrams representing the gap formula 22-4

1. We consider the applied pressure as given and obtain the flow from the gap formula. This corresponds to the Y representation.

2. With given flow, we use the gap formula to determine the necessary applied pressure, corresponding to the Z representation.

We shall need both ways or representations frequently in this book.

The average velocity of the volume flow in the gap is obtained by division with its cross section

$$v_m = \frac{1}{12\mu} \frac{P_0}{L} h^2 \qquad (22-5)$$

Finally, we determine the power consumed by the oil flow by multiplication with the applied pressure

$$|P| = \frac{1}{12\mu} \frac{P_0^2}{L} b\, h^3 \qquad (22-6)$$

This power increases with the square of the pressure as seen from equation 6.

We can draw the following conclusions from the gap formula equation 4:

FN[1] — Strictly speaking, fig 2 represents blocks as elements of functional diagrams as described in appendix 2.

1. The shear tension on the walls is evenly distributed over the length if the gap height remains constant. The pressure drop is, therefore, linear along the length and we obtain the pressure as shown in fig 3. In other words, there is a certain kind of symmetry since the pressure gradient, i.e. the derivative of the pressure along the gap length, does not depend on the position or on the value of the coordinate Z. The pressure gradient is, therefore, connected with the gap length and the applied pressure as follows

$$\frac{dp}{dz} = - \frac{p_0}{L}$$

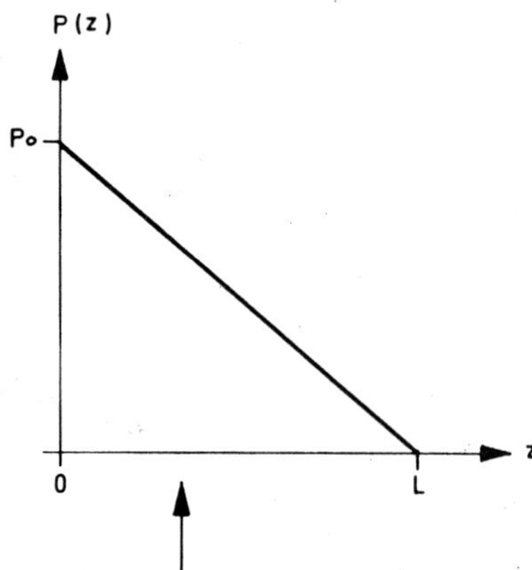


**Fig 22.3:** Distribution of pressure drop along the gap

If we insert the above form into equation 4, we obtain a differential form of the gap formula, /FN[2]/.

$$Q = - \frac{1}{12\mu} \frac{dp}{dz} b \, h^3 \qquad (22-7)$$

2. The linear pressure distribution represented in fig 3 determines also the forces which tend to push the walls apart. As we have seen in section 11, these forces are always obtained by integration of the pressure over the area of the walls.

---

FN[2] — Equation 7 can be derived more strictly from the Navier-Stokes equations of fluid mechanics.

$$F = b \int_{0}^{L} p(z)\, dz$$

With the linear pressure drop, the integral has the value $bLp_0/2$ as can be seen also from the area of the triangle in fig 3. Therefore, we obtain for the force pushing the walls apart

$$F = \frac{bL}{2} p_0 \qquad\qquad (22\text{--}8)$$

Formula 8 will be used, especially in section 33–4, for the calculation of the balancing of hydrostatic forces in machines and components. It should be noted that this force is independent of the gap height.

3. Apart from the forces themselves, it is important to know also their point of attack or their (tilting) moment on the wall. We know from geometry and can calculate quickly that the centre of gravity of a right-angled triangle is situated on one third of the height of a side as represented in fig 3 by the small arrow. The distributed force on the walls, according to equation 8, can therefore be replaced by a single force in that position, that is one third along the gap length. It produces a tilting moment depending on the reference point or the tilting axis. This is important for the suspension and moment balancing of some hydrostatic components as, for instance, cylinder blocks in axial piston machines.

4. Apart from the normal forces that tend to separate the walls, there exist the tangential forces that drag the walls in the direction of the pressure gradient. Their origin is the pressure force on the frontal face of the gap.

   The corresponding tangential forces are evenly distributed over both walls as can be seen either from the symmetry or by the calculation of the velocity gradients. Each wall experiences then a tangential force of

$$F_{ta} = \frac{b\,h}{2} p_0 \qquad\qquad (22\text{--}9)$$

5. The gap formula has been derived exactly in principle but one cannot expect great accuracy because of the assumed constant

viscosity. Actually, the oil temperature increases, diminishing the viscosity along the gap, as treated more in detail in section 24. Another disturbing factor is that even a small inclination or tilting of one wall with respect to the other strongly influences the gap height and thus the flow with given pressure.

### Numerical Example of The Gap Formula

As shown in fig 4, we have a piston with d = 2 cm diameter, a penetration length of L = 5 cm and a diametral clearance of 20 micrometers = $2 \cdot 10^{-3}$ cm in the cylinder. Viscosity $\mu = 0.2 \cdot 10^{-6}$ barsec and pressure $p_o$ = 200 bar.

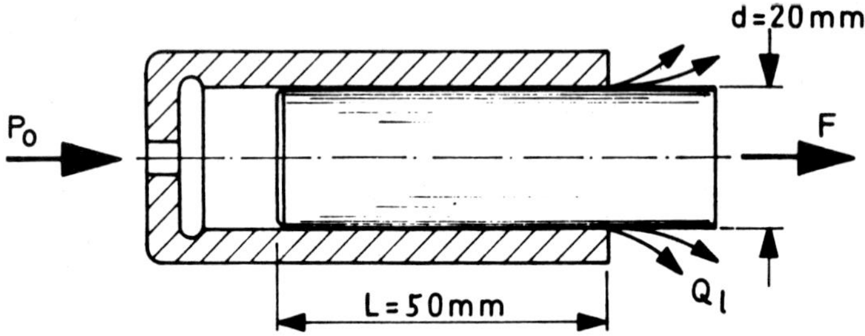


**Fig 22.4: Piston in cylinder as an example for the gap formula**

If the piston is located centrally and aligned with the cylinder, the gap height is everywhere the same and half the diametral clearance, thus h = $10^{-3}$ cm. The gap width is equal to the piston circumference of 6.28 cm.

The leakage flow Q becomes with formula 4

$$Q = \frac{200 \text{ bar}}{12 \; 2 \cdot 10^{-7} \text{ barsec}} \quad \frac{6.28 \text{ cm}}{5 \text{ cm}} \quad 10^{-9} \text{ cm}^3$$

$$Q = 0.105 \text{ cm}^3/\text{sec}$$

The cross section of the gap is equal to its circumference times gap height

$$A = 6.28 \cdot 10^{-3} \text{ cm}^2$$

giving a mean velocity of

$$v_m = \frac{Q}{A} = \frac{0.105}{6.28 \cdot 10^{-3}} \quad \frac{\text{cm}^3/\text{sec}}{\text{cm}^2} = 16.7 \text{ cm/sec}$$

In spite of the enormous pressure gradient, the mean velocity is rather small due to the strong braking by the shear forces and by wall adhesion.

With a different applied pressure, the leakage flow increases proportionally and, with a different gap height, proportionally to its third power so that it quickly becomes intolerable. The leakage with different applied pressures and gap height can be expressed by the following useful formula

$$Q = 0.105 \frac{cm^3}{sec} \frac{p_o}{200 \text{ bar}} \left[ \frac{h}{10\mu m} \right]^3 \qquad (22-4b)$$

## Further Remarks on the Gap Formula

The gap formula 4 forms an important basis of the design of hydrostatic machines, since it is applicable both to valve faces (section 33–4) and to pistons in cylinders and, furthermore, to many kinds of sliding valves. In the case of pistons, the width of the gap is simply the mean circumference of the piston and of the bore and the gap height equal to half the diametric clearance. On the other hand, it can be applied rigorously if the following two conditions are met:

1. The piston must be located centrally in the cylinder.

2. The piston axis must be parallel to the cylinder axis.

If the axis of piston and cylinders are not parallel, we obtain a variation of the gap height in the direction of the pressure gradient, as treated in section 23. On the other hand, we shall examine now a piston located eccentrically in the cylinder and also the case of the gap of finite  i.e. not very large, width.

## Eccentric Pistons

When calculating the leakage of an eccentric but parallel piston, we suppose first that the piston is entirely eccentric contacting the cylinder wall in a line. The gap height is then variable around the circumference between zero and twice the radial clearance and its third power between zero and eight times the value with a central piston. We expect therefore an increase of leakage, as easily confirmed by a detailed calculation. This is not reproduced here, but we note that, with entirely eccentric position, the leakage is increased by a factor of 2.5 as compared to the simple gap formula 4 with half diametric clearance as gap height. With smaller eccentricities, the increase of the leakage is correspondingly smaller.

In order to determine the leakage with arbitrary eccentricity, one can split the gap into several leakage paths, each having a different gap height, but constant along the z coordinate, as determined from the geometrical relations of two circles with a small difference of diameters. The leakage flow is obtained by summation of all leakage paths, i.e. by integration around the circumference. Starting from its value with central pistons, the leakage increases steadily with increasing eccentricity until reaching the value of 2.5 with the piston touching the wall.

Further discussion of eccentrical pistons is outside the scope of this book and the interested reader is referred to the book of Guillon, B.7, appendix 1.5, page 34.

### Gaps with Small Width

Our gap formula 4, derived for infinitely wide gaps, is valid only if the gap width is large compared to the gap height. Reducing the gap width until it becomes comparable or, finally, equal to the gap height produces an additional braking of the flow due to the shear forces on the side walls. Consequently, there will be less flow with a given applied pressure and this can be represented by multiplying the gap formula 4 by a function, smaller than 1, of the ratio of gap width to gap height.

$$Q = \frac{1}{12\mu} \frac{p_0}{L} b \; h^3 \; f\left(\frac{b}{h}\right) \tag{22-10}$$

We know from the infinitely wide case that the function f has the value of 1 for very large arguments and we expect that it decreases steadily with decreasing argument. The calculation of the velocity field with finite gap width is a problem of potential theory and has infinite Fourier series as solution. The flow is then obtained by integration over the velocity field and conveniently represented as in equation 10.

The function f of equation 10 is represented graphically in fig 5. We see that, with a ratio of b/h = 10. there is only 6% reduction of flow as compared to the infinitely wide case and a square gap has still 42% of the original flow. We can, therefore, establish the practical rule that the influence of the side walls is negligible for a width/height ratio of more than 10.

The table still contains a last entry 'circular' corresponding to the small circle at the right of fig 5. It represents the leakage of a round gap of the diameter h, equal to the gap height. We see that the flow in the circular case is only 29% of the infinitely wide gap of the same height, still considerably reduced compared to the square gap. This should be expected since the corners between the circle and the square are no longer available for the flow. Incidentally, the circular case represents the well known Haagen-Poiseuille formula.

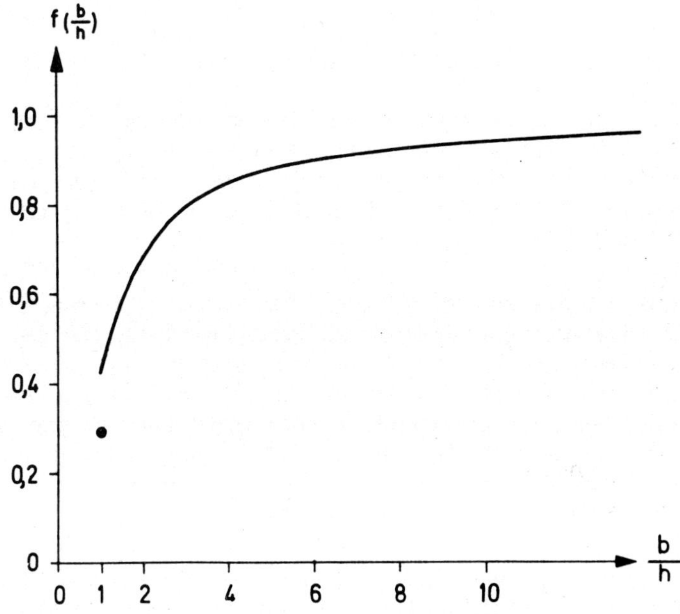

Fig 22.5:Reduction of the flow with not so wide gaps, according to equation 22-10

As a final remark, the gap around a piston in a cylinder is rolled up and the braking side walls are missing. Consequently, the formula for the infinitely wide gap is applicable.

## 23    Gaps with Variable Height

### 23—1    Stationary Walls

In section 22, we have studied the gap with parallel stationary walls and derived the gap formula 22—4 for flow and applied pressure. We shall examine now how the flow behaves if the gap height is variable in the direction of the pressure gradient, first with stationary and further below with moving walls.

Gaps with variable height occur frequently, for instance due to slight misalignment of hydrostatic bearing members. The gap walls remain nearly parallel in all cases of practical importance or, in other words, the angle between the walls is small, only a few milli-radiants. Since the gap length is much larger than the gap height, this does not prevent large variations of the gap height, for instance in the range of one to two. For simplicity, we shall neglect the corrections for finite gap width in formula 22—8.

The applied pressure drives the oil along the walls. Due to the small inclination of the walls, the velocity component across is much smaller than the component along the gap and can be neglected without hesitation. It is now more convenient to consider the flow as given and to calculate the pressure drop, /FN¹/. We shall, therefore, use the gap formula in the Y representation as on the right of fig 22—2. Furthermore, since the oil is incompressible, the volume flow is constant along the gap.

With the small inclination of the gap walls, each element of length is governed by equation 22—7 in spite of the fact that the gap height is now a function of the co-ordinate z along the gap. This allows to solve equation 22—7 for the pressure gradient

$$\frac{dp}{dz} = - \frac{12 \mu Q}{b\, h^3} \tag{23-1}$$

FN¹ — Amusing remark from the lecture: Technical and scientific investigations of this kind are like a game of sport; it is important to outwit nature and to assume the most favourable position before really starting.

With a given flow and with the gap height given as function of z, it is easy to compute the pressure build-up by integration of equation 1. This includes the pressure at the entrance of the gap. We note immediately from formula 1 that all pressures are proportional to the flow since this is true for the pressure gradient.

If the applied pressure rather than the flow is given — as usual in practice —, the flow can be calculated easily by inversion, i.e. by solving of the calculated relation for the pressure.

Let us consider now two important cases of variation of gap height:

a) *Linear Decrease of Gap Height*

The most simple geometry of a gap with variable height is obtained if the gap height decreases *linearly* over the gap length from a beginning value $h_1$ to a final value $h_0$. Fig 1 shows such a decreasing gap where the oil enters from the left and flows out to the right. Behind the gap, there is no pressure since the outflow of the oil is not impeded or throttled. With the imposed flow, pressure develops in the gap as we shall calculate now.

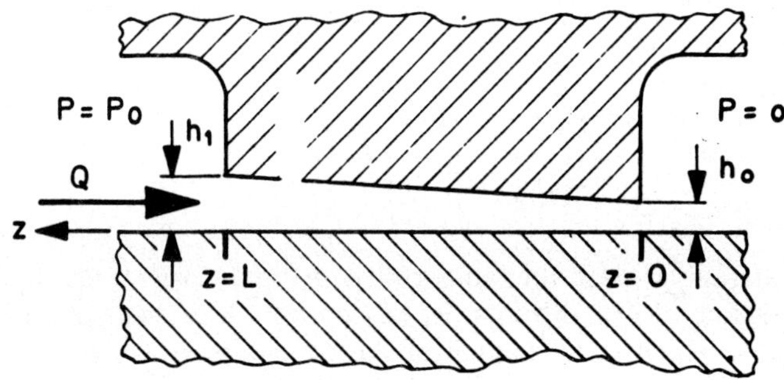


Fig 23.1:    Cross-section through a gap with reducing gap height

It is convenient to change the sign of the z co-ordinate in such a way that it runs now from the right to the left, as shown in fig 1, allowing integration from z = O until z = L. Consequently, the minus sign of equation 1 must be removed

$$\frac{dp}{dz} = \frac{12 \mu Q}{b\, h^3}$$

The above equation shows how the pressure builds up beginning from the end of the gap. We obtain then the pressure by integration

$$p(z) = \frac{12\mu Q}{b} \int_{0}^{z} \frac{dz}{h^3(z)} \tag{23–2}$$

The linearly reducing gap height can be expressed as follows

$$h = h_0\left[1 + \delta \frac{z}{L}\right] \quad \text{and} \quad h_1 = h_0 + h_0\,\delta \tag{23–3}$$

Here, $\delta$ is a dimensionless variable describing the slope of the gap. As we can see from the second form of equation 3, $\delta$ gives the ratio between the height at the beginning and at the end of the gap. If, in particular, $\delta = 1$, then we have $h_1 = 2h_0$.

Inserting equation 3 into equation 2, we obtain

$$p(z) = \frac{12\mu Q}{b} \int_{0}^{z} \frac{dz}{h_0^3\left(1 + \delta \frac{z}{L}\right)^3} \tag{23–4}$$

Substituting $u = \delta\, z/l$

$$p(z) = \frac{12\mu Q L}{b\, h_0^3\, \delta} \int_{0}^{u} \frac{du}{1 + u^3}$$

Equation 4 describes the pressure development as a product expression of the different operating variables multiplied by the integral, which is a function of the dimensionless upper limit. We are interested especially in the applied pressure at the entrance of the gap obtained with $z = L$ or if $u = \delta$. The integration itself presents no difficulty and the result is best represented as follows

$$p_0 = \frac{12\mu Q L}{b \cdot h_0^3} \frac{1}{f(\delta)} \quad \text{with} \quad f(\delta) = \frac{2(1 + \delta)^2}{2 + \delta} \tag{23–5}$$

The Formula 5 shows the pressure build up with given operating variables and especially volume flow. In order to determine the flow with given pressure, we invert equation 5, /FN²/ and, at the same time, replace $1 + \delta$ by $h_1/h_0$ leading to

FN² ; That means simply to solve it for Q.

$$Q = \frac{p_o\, b\, h_o^3}{12\, L}\; f\left(\frac{h_1}{h_o}\right) = Q_\infty \cdot f\left(\frac{h_1}{h_o}\right) \qquad (23\text{–}6)$$

In equation 6, $Q_\infty$ denotes the flow in a gap with constant height $h_o$. The function of the ratio $h_1/h_o$ indicates the increase of the flow by the inclination of the walls, as represented in fig2. It is seen that, with a height ratio of $h_1/h_o = 2$, the flow is increased by a factor 2.7.

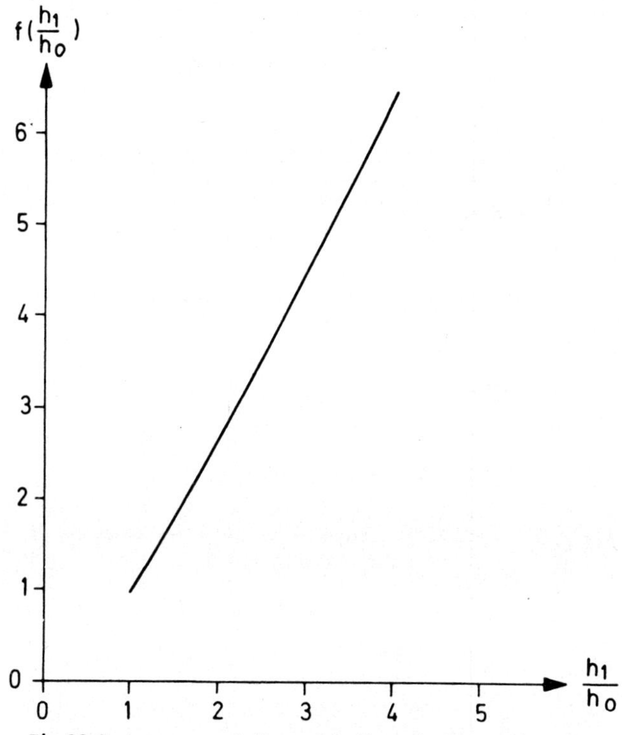


Fig 23.2: Increase of flow with linearly increasing gap
height, according to equation 23-6

The angle of the walls is very small, even if the gap height changes greatly between entrance and exit. As an example, with $h = 10\ \mu m$ and $L = 5\,cm$, the gap height doubles with an inclination of $10\,\mu m$ on a length of 5 cm corresponding to an angle of 0.2 milli-radiants.

We shall use the pressure development as discussed here for the stability of pistons and valve spools.

## b) *Parabolic Gap Height*

A further important form of the gap is produced by a flat, plane surface and by a cylinder almost in contact. A spherical piston in a cylinder contains this form rolled up around a longitudinal axis. It is represented in fig 3 with the z co-ordinate running again from right to left. The origin of the z co-ordinate is located at the point of nearest approach, i.e. on the equator of the sphere. In this region, the circular cross section of the sphere can be approximated by a parabola, as known from analytic geometry. We obtain therefore the following equation for the gap height as a function of the co-ordinate z with R = radius of cylinder

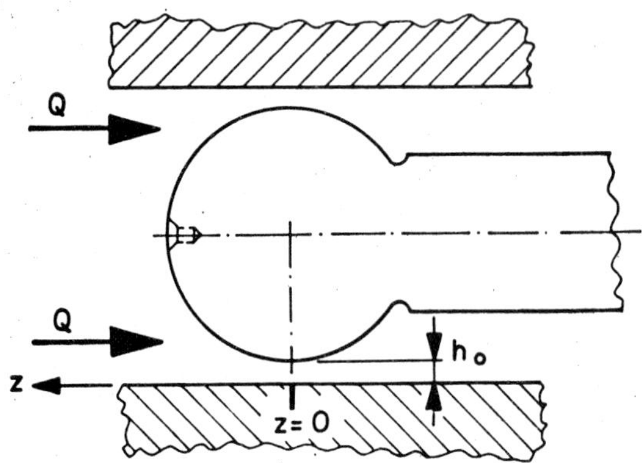


**Fig 23.3:**   Spherical  piston  in  cylinder  as  example  of
parabolic gap height

$$h\,(z) \;=\; h_o \;+\; \frac{z^2}{2R}$$

Considering again the flow as given and computing the necessary applied pressure, we insert this gap height into formula 2. The determination of the lower limit of integration can be extended to infinity $(-\infty)$, since the pressure drop is concentrated around the region of the smallest gap height because the third power of the gap height increases and its reciprocal decreases rapidly away from the equator. As an example, if the gap height has increased by a factor of 3, the pressure gradient has fallen to 1/27. In general, the pressure gradient becomes negligible long before the parabolic approximation becomes inaccurate.

We obtain now for the pressure as a function of the length

$$p(z') = \frac{12\,\mu\,Q}{b} \int_{-\infty}^{z} \frac{dz}{h_o^3 \left[1 + \dfrac{z^2}{2\,h_o\,R}\right]}$$

Substituting the dimensionless expression $u^2 = z^2/2h_oR$, we obtain

$$p(z) = \frac{12\,\mu\,Q\,\sqrt{W2h_oR}}{b\,h_o^3} \int_{-\infty}^{u} \frac{du}{(1+u^2)^3} \qquad (23\text{–}7)$$

This pressure development is again a product expression with the operating variables multiplied by the dimensionless integral a function of the upper limit of the integral.

The integral of formula 7 can be determined from a table of integrals or by decomposition into partial fractions. We note here the result of integration for completeness

$$\int_{-\infty}^{u} \frac{du}{(1+u^2)^3} = \frac{u}{4\,(1+u^2)^2} + \frac{3}{8}\,\frac{u}{1+u^2} + \frac{3}{8}\,\tan^{-1}u$$

We are mainly interested in the pressure that is built up at the entrance, i.e. the applied pressure far above the equator. The upper limit of integration can be extended to infinity for the same reason giving $3\pi/8$ for the integral

$$p_o = \frac{12\,\mu\,Q}{b\,h_o^3}\,\sqrt{h_oR}\qquad 1.67$$

With a given pressure, we obtain the flow between a spherical piston and a cylinder by inversion of the above formula. In analogy with the gap formula, the result is best represented as follows

$$Q = \frac{p_o\,b\,h_o^3}{12\,L_{eq}} \qquad\qquad L_{eq} = 1.67\,\sqrt{R\,h_o} \qquad (23\text{–}8)$$

Here $L_{eq}$ is an equivalent length, very useful for comparing the leakage of a spherical piston with the leakage of a cylindrical piston with the same diameter and diametric clearance.

### Numerical Example of Parabolic Gap Height

Consider a spherical piston with the same dimensions as in section 22, i.e. $h_o = 10^{-3}$ cm, R = 1 cm. The equivalent length, also called sealing length, becomes

$$L_{eq} = 1.67 \sqrt{10^{-3} \text{ cm} \cdot 1 \text{ cm}} = 5.25 \cdot 10^{-2} \text{ cm}$$

The sealing length is therefore only about 2.5% of the piston length or only 1% of the length of the example in section 22. Therefore, it is necessary to use much tighter clearances with spherical pistons.

We see from our numerical example that a spherical piston is much less efficient for sealing than a cylindrical one. Therefore, one manufacturer of axial machines with spherical pistons improves sealing by piston rings. We note also from formula 8 that the equivalent length becomes smaller with reduced diametric clearance.


## 23–2     Moving Walls

Let us now consider the general case of a gap with moving walls and variable gap height and with an external pressure applied to the gap. We assume for the moment only a movement of the wall in the direction of the pressure gradient whilst any perpendicular movement will be considered below.

The simplest case of a moving wall with constant gap height and without applied pressure has been already treated in section 21. There, the gap of fig 21–1 delivers oil to the right without producing pressure within the gap.

The basic arrangement of the gap with moving walls and variable height is similar as in fig 1, only the lower wall or support is moving now with the velocity v to the right and the gap height is an arbitrary function of the z co-ordinate. The angle of inclination of the top wall is again very small.

The oil adheres, due to the viscosity, to both walls and is drawn into the gap by the movement of the lower wall. Since the gap narrows on fig 1, there is a kind of crowding effect producing a pressure field. It pushes the walls apart, as we know from section 11—4, and this effect is called hydrodynamic lubrication (see section 26).

It is instructive to consider what happens if the gap height increases in the direction of movement. The oil flow drawn in at the entrance is then no longer sufficient to fill up the higher gap and cavities with vapour or air They form a kind of pattern like the fingers of a hand, /FN [3/].

The main problem is now to determine the pressure development in the general case with moving walls, variable gap height and externally applied pressure. As before, the volume flow over the gap length remains constant since the oil is considered incompressible and the gap infinitely wide.

Our general case is covered by the following two particular cases:

1.  With stationary walls and variable gap height, the relation of flow and pressure gradient follows formula 22—7.

2.  With constant gap height and one moving wall, no pressure builds up within the gap and the flow is given by formula 21—2.

We can now simply add the flows of both particular cases 1 and 2, i.e. equations 22—7 and 21—2, to determine the equation in the general case

$$Q = \frac{v_0\, h\, b}{2} - \frac{b\, h^3}{12\, \mu}\, \frac{dp}{dz} \qquad (23–9)$$

Whilst this addition of the flows appears reasonable, it can be derived rigorously from the Navier-Stokes equations. It is possible since the relation between flow, pressure gradient and wall movement is linear.

When using equation 9, it is to be noted that the positive z axis is directed to the right. The flow Q on the left side of equation 9 remains constant and, with variable height, the pressure gradient develops exactly so that this flow condition is fulfilled. In other words, the total flow consists of the entrainment flow according to equation 21—2 and of the flow due to the pressure gradient according to equation 22-7.

---

FN[3] — L. Floberg, 'Cavitation in Lubricating Oil Films', in the book R. Davis, Editor, 'Cavitation in Real Liquids', Elsevier, 1964.

For the solution of equation 9, it is usually convenient to consider the flow as given and to determine the pressure gradient and the pressure itself by integration. This gives the relation between flow and applied pressure or the characteristics of our gap. If the applied pressure is given, we obtain the flow again by inversion of the characteristics. It should also be noted that, even without applied pressure, a flow is delivered to the right, and pressure fields are built up inside the gap.

Examining now the case of a wall movement perpendicular or oblique to the pressure gradient, the perpendicular velocity component of the wall movement has no influence on the oil pressure since it entrains only the oil out of the plane of the paper. The only effect of this component is an increase of the power dissipated and of the temperature in the oil film, as we shall treat in section 24.

## 24      Thermal and Inertia Effects in Gaps

### 24—1      Thermal Effects

So far, we have excluded the inertia effects or the forces necessary to accelerate the liquid in the gap and we have assumed constant viscosity. This is a very restrictive condition and does not
correspond well to the practice of hydraulics since the oil heats up
on its flow through the gap. We shall, therefore, examine here the
temperatures and below the inertia effects assuming constant gap
height for simplicity.

In order to determine the temperatures, it is convenient to set up
a balance of the different powers entering or leaving the gap. This is
represented in fig 1 with an enlarged representation of a section of
the gap. We have there the following power flows:

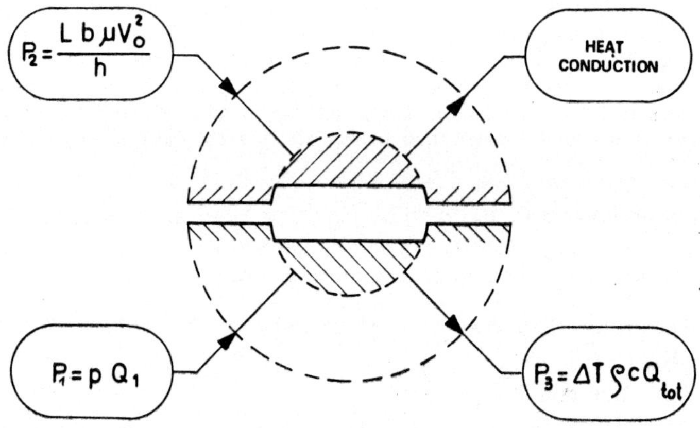


Fig. 24.1: Symbolic   representation   of   power   balance
in   gaps

1. The power $P_1$ of the leakage flow with stationary walls and of
   the applied pressure.
2. The power $P_2$ due to shearing of the oil film according to
   formula 21—5.

3. The power $P_3$ in the form of heat or temperature increase of the entire flow leaving the gap.

4. The power $P_4$ conducted away through the walls as heat.

The first two power components enter the gap element where they are transformed into heat and removed by the third and fourth power flow, and the balance determines the temperatures.

One might have doubts in the general case with applied pressure and with moving walls about the addition of the powers due to the oil flow with stationary walls and due to shearing with moving walls. However, this is admissible, as can be shown by the following considerations. Assuming first a wall velocity component only across the pressure gradient or perpendicularly to the plane of the paper in our figures, it produces a corresponding flow velocity field. Since the power dissipation is quadratic in the wall velocity according to equation 21—5, its contribution can be added by the theorem of Pythagoras to the contribution of the longitudinal velocity field.

Any component of the wall velocity parallel to the pressure gradient influences the leakage flow according to equation 23—9 and the power supplied by the applied pressure and leakage flow changes by equation 22—6. On the other hand, due to the pressure gradient, a force acts on the wall after equation 22—8. This represents a power together with the wall velocity and a calculation shows that these power changes are equal and annul each other. Assuming as an example a pressure applied in fig 1 from the left, then the wall velocity increases the flow and the power drawn into the gap, but the pressure gradient drives the wall to the right removing the same power from the oil film. Summarising, we have shown why it is admissible to determine the total power flow in the oil film by addition of the single powers.

Whilst the power $P_1$ is given by equation 22—6 and the power $P_2$ by equation 21—5, we obtain the power $P_3$ carried by the increased temperature of the outgoing flow from the product of:

1. The total flow due to wall movement and pressure gradient $Q_{tot}$.
2. The mass density $\rho$.
3. The specific heat $c_v$, i.e. h the ratio of heat energy and mass.
4. The temperature increase $\Delta T$.
5. The heat equivalent J.

Since heat is a form of energy, the heat equivalent is a dimensionless conversion factor in the sense of the dimensional theory of physical variables, as treated in appendix 3. It becomes equal to one if one uses the joule instead of the calory as the unit of heat.

Coming to $P_4$, the power flow due to heat conduction of the walls, it is very difficult to say anything in general since it depends too much on the overall design and of the cooling of the adjacent parts. In other words, this power is seen from the gap largely undetermined.

We can, therefore, set up the following equation

$$P_1 = \frac{p_0^2}{12\mu} \frac{b}{L} h^3 = p_0 \, Q_1$$

$$P_2 = b L \mu \frac{v_0^2}{h}$$

$$\tag{24-1}$$

$$P_3 = Q_{tot} \, \rho \, c_v \, \Delta T \, J$$

$$P_4 = \text{indeterminate}$$

In many cases of practical importance, the powers $P_2$ and $P_4$ are negligibly small. If we can, furthermore, neglect the flow due to the wall movement, then we can determine the temperature increase of the leaving flow by comparison of the remaining powers $P_1$ and $P_3$ as follows

$$\Delta T = \frac{p_0}{c_v J \rho} \tag{24-2}$$

*Numerical example* of the important formula 2 with an applied pressure of 100 bar. With the values:

$$\rho = \text{mass density} = 0.9 \text{ gr/cm} = 0.9 \cdot 10^{-3} \text{ kg/cm}^3$$

$$J = \text{heat equivalent} = 4.18 \frac{J}{cal} = 41.8 \frac{daNcm}{cal}$$

$$c_v = \text{specific heat} = 0.43 \text{ cal/gr}^\circ\text{C},$$

we obtain

$$\Delta T = 100 \frac{daN}{cm^2} \frac{1}{0.43} \frac{gr^\circ C}{cal} \frac{1}{41.8} \frac{caL}{daNcm} \frac{1}{0.9} \frac{cm^3}{gr} = 5.5^\circ C$$

Since the temperature increases proportionally to applied pressure, we can write for different applied pressures

$$\frac{\Delta T}{5.5^\circ C} = \frac{p_0}{100 \text{ bar}} \qquad (24\text{--}2b)$$

The influence of the wall movement on the temperature depends again on the fact whether the movement is parallel or perpendicular to the pressure gradient. A wall movement perpendicular to the pressure gradient has no influence on the temperature if the gap is infinitely wide as long as $P_2$ is negligible because it does not influence the outflow of the oil. With a finite gap width, the effect depends on the temperature difference of the flows entering and leaving on the side. A gap closed on the side in itself, like the clearance between pistons and cylinders or the sealing lips on the distributing surfaces of axial piston machines (section 33–3), is infinitely large in this sense.

A wall movement along the pressure gradient influences the flow and the temperatures. As an example, if the wall is moving in the direction of the pressure gradient, the flow increases, leading to less temperature increase. With a reverse movement, the flow decreases and may even vanish so that less power is removed by it. This case can produce dangerous temperatures and local hot spots leading eventually, with insufficient heat conduction of the walls, to seizing and destruction of the walls.

With rapid lateral movement of the wall of the infinitely wide gap, the corresponding power component becomes appreciable, allowing us to represent the entire power flow as follows

$$P_1 + P_2 = P_1 \left[ 1 + 12 \frac{L^2 \mu^2 v_0^2}{p_0^2 h^4} \right] \qquad (24\text{--}3)$$

There is a corresponding increase of the temperature of the oil leaving the gap which we shall write, in analogy to the numerical example of equation 2, as follows

$$\frac{\Delta T}{5.5^\circ C} = \frac{p_0}{100 \text{ bar}} \left[ 1 + 12 \frac{L^2 \mu^2 v_0^2}{p_0^2 h^4} \right] \qquad (24\text{--}4)$$

*Numerical example:*     In order to get a feeling for the orders of magnitude, we determine the additional heating due to wall motion with the following data:

$$L = 0.5 \text{ cm}$$
$$\mu = 2 \cdot 10^7 \text{ barsec}$$
$$v = 500 \text{ cm/sec}$$
$$p = 100 \text{ bar}$$
$$h = 10^{-3} \text{ cm}$$

We obtain

$$\frac{P_1 + P_2}{P_1} = 1 + 12 \left[ \frac{L \mu v_c}{p_0 h^2} \right]^2$$

$$\frac{P_1 + P_2}{P_1} = 1 + 12 \left[ \frac{0.5 \text{ cm} \cdot 2 \cdot 10^{-7} \text{ barsec} \cdot 5 \cdot 10^2 \text{ cm/sec}}{100 \text{ bar} \qquad 10^{-6} \text{ cm}^2} \right]$$

$$= 1 + 12 \left[ 0.5 \right]^2 = 4$$

The above numerical values produce a fourfold increase of temperature, corresponding to 22°C with 100 bar applied pressure.

The most important consequence of equation 4 is that the temperature of the outflowing oil increases very rapidly with smaller gap height as long as heat conduction of the wall remains negligible. It is one of the reasons determining the minimum gap height for safe operation of many hydrostatic machines, as we shall treat in section 25–6.


## 24–2     Inertia Effects

So far, we have neglected the effects of inertia or of the mass of the liquid in the gap. We shall now consider its influence assuming a given flow through a gap with parallel stationary walls. We would like to determine the pressure needed to accelerate the liquid to the velocity distribution described by equation 22–3. The calculations can proceed as in section 13, only we have now to integrate over each flow path according to its velocity as given by equation 22–3.

We expect that the power needed will be more than with an even velocity distribution and it turns out that we need just twice the dynamic pressure. This allows to write for the pressure drop on the gap including inertia forces

$$p \; = \; \frac{12\,\mu\,Q\,L}{b\,h^3} \; + \; \rho \left[\frac{Q}{b\,h}\right]^2 \tag{24-5}$$

Equation 4 is represented as a functional diagram in fig 2, show-ing how the viscous and the inertia pressure drop are added with a prescribed flow. If the applied pressure is given, it is best to cal-culate first the flow according to the gap formula 22—4 and then to determine the inertia pressure drop as a correction, as shown in fig 3 as a functional diagram. The additional pressure drop is usually negligibly small with the low velocities in gaps. In the numerical example in section 22, the mean velocity was 17 cm/sec and to this corresponds a pressure drop of less than 1 millibar.

When using the gap formula in order to determine the viscosity, normally in a round narrow tube or capillary, the additional pressure drop due to inertia is known as Hagenbach's correction factor.

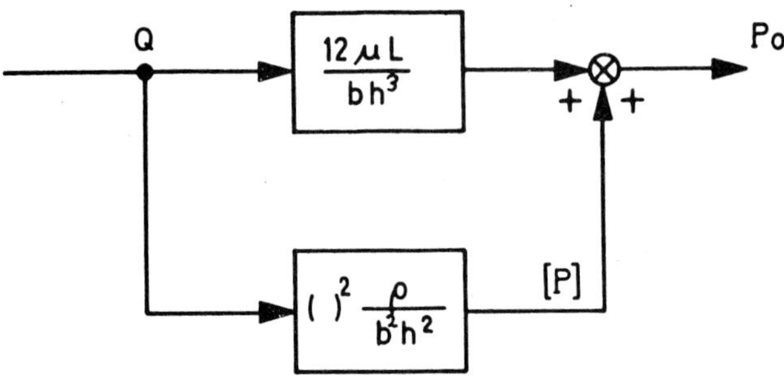



Fig 24.2: Functional diagram for inertia forces with flow as input variable

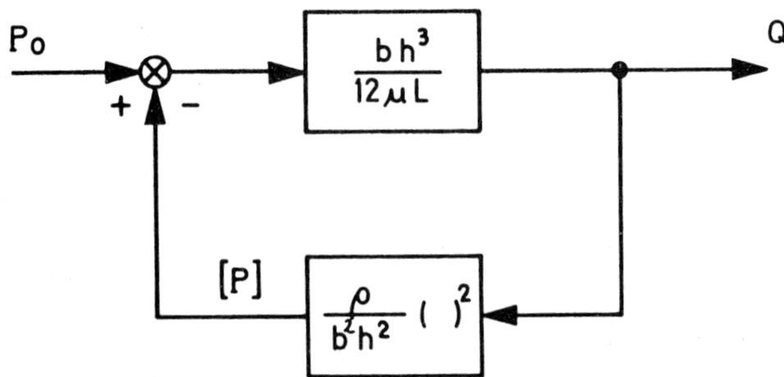



Fig 24.3: Functional diagram for inertia forces with pressure as input variable

# 2 THE GAP AS BASIC HYDROSTATIC ELEMENT

## 25        Hydrostatic Bearings

### 25–1        The Slipper as Basic Form

Hydrostatic bearings are especially important in hydrostatic engineering because the working fluid can be used immediately for lubrication. They have many different forms but the essential features are all represented in the simple slipper.

The upper part of fig 1 shows a cross section through a slipper loaded with the force $F_{ld}$, over its spherical head, as shown. This spherical head allows a self-alignment of the angular position of the slipper relative to the support unless disturbed by too much friction. Therefore, we introduce a lubrication duct as indicated or apply an antifriction treatment, for instance PTFE, on the spherical surfaces. On the bottom of the slipper, we find a recess or depression with the area $A_{cv}$ surrounded by sealing lips with the area $A_{lp}$ The surface of the lips is plane or flat and constitutes, together with the plane support, a gap with an oil film.

The lubricant is supplied from a pressure source with a supply pressure $p_{sp}$ over a throttle where the pressure reduces to $p_o$, the pressure applied to the sealing lips. The pressure drop on the throttle depends on the flow according to its impedance and, therefore, on the gap height, i.e. the distance between the sealing lips and the support.

The graph in the lower part of fig 1 shows the pressure development with its trapezoid form under normal load in continuous line. The pressure in the depression is determined by the impedances of the throttle and of the sealing lips. If the load is increased by a perturbation, the gap height is reduced and the flow and the pressure drop in the throttle decrease. Therefore, the applied pressure $p_o$ increases until it compensates the increased load. In the limiting case,

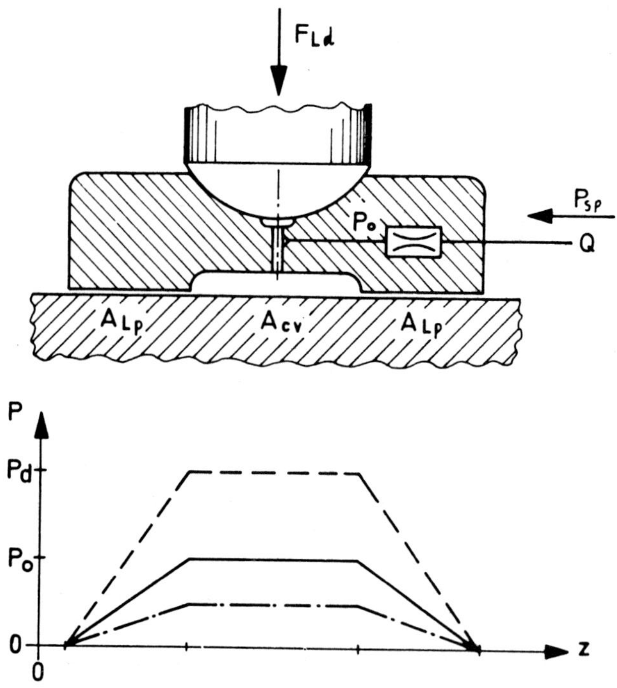



Fig. 25.1: Slipper as basic form of a hydrostatic bearing
with throttle and transmission of load forces by
spherical seat

if the walls close up completely, the applied pressure increases up to
the supply pressure. Conversely, the applied pressure decreases on
unloading, leading to the point-broken pressure distribution with increasing gap height. In the limiting case with vanishing load, it
becomes zero, when the entire supply pressure is dissipated by
the throttle.

We see, therefore, how the slipper can support perturbation forces
without metallic contact and without complete separation. This is
the important stability of the hydrostatic bearings depending on the
change of the pressure drop on the throttle. We see that the throttle
is essentially important for correct operation if the slipper is fed
from a pressure source.

If the slipper is supplied from a flow source, on the other hand,
where the flow is independent of the supply pressure, no throttle is
needed. The operation is generally the same, only, with decreasing

gap height, the applied pressure and the force increase without limit. Hydrostatic bearings with a flow source are used sometimes in the design of very large machines where the special lubricant pump can be accepted.

We shall introduce the *load factor* $\lambda$ as important characteristic parameter for hydrostatic bearings supplied by a pressure source. It is simply the ratio of supply to applied pressure at normal load.

$$\lambda = \frac{p_o}{p_{sp}}$$

The load factor can thus vary between 0 and 1. Increasing this factor increases the load that a slipper can carry with given dimensions and supply pressure but decreases the stability. In the limit of $\lambda = 1$, there is no stability or the slipper carries a load independently of gap height since there is always the supply pressure in the depression. In many practical cases, a load factor of $\lambda = 0.5$ is a good choice.

We shall now calculate the load carried under the following assumptions:

   1.  The slipper is parallel to its support and the gap height constant.

   2.  The viscosity is constant across the sealing lips.

With these assumptions, we obtain a linear pressure drop over the lips, as treated in section 22, independently of the gap height and the load force becomes

$$F_{ld} = p_o \left[ A_{cv} + \frac{A_{lp}}{2} \right] = \lambda \, p_{sp} \left[ A_{cv} + \frac{A_{lp}}{2} \right] \qquad (25-1)$$

which can be written as

$$F_{ld} = p_o \, A_{tot} \qquad A_{tot} = A_{cv} + \frac{A_{lp}}{2} \qquad (25-1a)$$

The flow or lubricant consumption is now given by the gap formula 22-4, using as width b the sum of the mean width of all lips. It is valid also if the slipper moves on its support, because the flow drawn in at the front sealing lip by equation 23-9 leaves again through the rear sealing lip, if the gap height is constant. The flow over the lateral sealing lips is not influenced by the slipper movement.

Any tangential forces tend to impede the movement of the slipper on its support. However, the tangential forces produced by the pressure drop after equation 22-9 annihilate themselves on the circumference of the slipper, since they act equally in one as in another direction, leaving only the tangential forces due to the relative movement of the walls. These forces and the corresponding power consumption have been treated in equation 21—1 and 21—2 which we write as follows with the lip area $A_{lp}$

$$F_{ta} = \mu \, A_{lp} \, \frac{v_o}{h} \qquad\qquad P = \mu \, A_{lp} \, \frac{v_o^2}{h} \qquad\qquad (25\text{--}2)$$

We see from equation 2 that the tangential forces and the corresponding power loss depend only on the area of lips and not on their form.

For comparison with other bearings, it is interesting to calculate the coefficient of friction or the ratio of the tangential to the load force. We obtain by combining equations 2 and 3 and by the expression $v_o = \omega r$, where r is a radius (mean radius in the case of axial bearings),

$$f_{red} = \frac{F_{ta}}{F_{ld}} = \frac{2 \mu \omega}{p_o} \, \frac{r}{h} \, \frac{1}{1 + 2 \, A_{cv}/A_{lp}} \qquad\qquad (25\text{--}3)$$

The total power consumption is given by the consumption due to the tangential forces above and by the consumption due to the fluid flow. For a fair comparison with other bearings, the power loss of the throttle, essential for stability, should be included. Hence, the power consumed by the fluid flow becomes simply the product of supply pressure and flow

$$P = Q \, p_{sp} \qquad\qquad (25\text{--}4)$$

It is also possible to express the power consumption of the fluid flow as an equivalent additional coefficient of friction. However, in hydrostatic machines, the difference of characteristics should be kept in mind. Particularly, the fluid supply requires much power at high pressures and low velocities. Finally, it is interesting to note how equation 3 is composed with the dimensionless product expression $\omega\mu/p_o$ appearing again in Section 32 in the mathematical models of hydrostatic machines. It is also used in the theory of hydrodynamic bearings (Section 26) with the mean bearing pressure produced by viscosity action in the narrowing gap. Equation 3 contains furthermore the geometrical form factor r/h, usually very large (one thousand and more), whilst the form factor with the areas is nearer to but less than one, usually around one third.

### Design Example of a Slipper

Let us calculate a slipper with $d_e = 32$ mm having to support the force of a piston with a diameter of 24 mm and an area $A_p = 4.52$ cm. Such a slipper is used in swashplate axial piston machines as described in Section 33–3 and the external diameter is limited by fouling of adjacent slippers.

We chose a length of 4 mm for the sealing lip corresponding to a diameter of the depression of 24 mm. Therefore, we obtain a mean diameter of $d_m = 28$ mm $= 2.8$ cm and $A_{lp} = 3.5$ cm$^2$, $A_{cv} = 4.52$ cm$^2$ .

The load factor becomes with

$$A'_p \cdot p_{sp} = p_o \left[ \frac{A_{lp}}{2} + A_{cv} \right]$$

$$\lambda = \frac{p_o}{p_{sp}} = \frac{A_p}{A_{lp}/2 + A_{cv}} = \frac{4.52}{6.2} = 0.73$$

about 73%, a little high but still acceptable. With an operating or supply pressure of 100 bar, we have an applied pressure of 73 bar in the depression, whilst 27 bar are dissipated by the throttle.

The oil flow becomes with 100 bar, $h = 12\mu m = 1.2 \cdot 10^{-3}$ cm

$$Q = \frac{1}{12} \frac{73 \text{ bar}}{2. \, 10^{-7} \text{ bar sec}} \frac{8.8 \text{ cm}}{0.4 \text{ cm}} (1.2 \cdot 10^{-3} \text{ cm})^3 = 1.15 \text{ cm}^3/\text{sec}$$

needing a throttle with a secant impedance of

$$Y = \frac{1.15 \text{ cm}^3/\text{sec}}{27 \text{ bar}} = 0.042 \frac{\text{cm}^3/\text{sec}}{\text{bar}}$$

The example of an axial piston pump in Section 32 with a displacement of 19 cm$^3$/rd has a pitch radius of the cylinders of 48 mm leading, with a rotation frequency of 157 rd/sec, to a sliding velocity of 720 cm/sec. We calculate first the coefficient of friction by equation 3 for an operating pressure of 100 bar

$$\mu\omega = 2 \cdot 10^{-7} \cdot 157 \text{ bar} = 3.14 \cdot 10^{-5} \text{ bar}$$

$$\mu\omega/p_o = \frac{3.14 \cdot 10^{-5}}{73} = 4.3 \cdot 10^{-7} \quad \text{since } p_o = 73 \text{ bar}$$

$$A_{cv}/A_{lp} = 4.52/3.5 = 1.29$$

$$f_{red} = 2 \cdot 4.3 \cdot 10^{-7} \frac{4.8 \text{ cm}}{1.2 \cdot 10^{-3} \text{ cm}} \frac{1}{1 + 2 \cdot 1.29} = 0.98 \cdot 10^{-3}$$

$$(25-3a)$$

We obtain thus a coefficient of friction of only one permill (or tenth of percent).

The tangential force can also be calculated directly by equation 2

$$F_{ta} = 2 \cdot 10^{-7}\ \text{barsec}\ \frac{3.52\ \text{cm}^2}{1.2 \cdot 10^{-3}\text{cm}}\ 720\ \frac{\text{cm}}{\text{sec}} = 0.42\ \text{daN}$$

All nine slippers of this pump contribute therefore a drag force of 3.8 daN and a corresponding torque of 18.3 daNcm (and a loss pressure of 1 bar, see section 32−3).

**25−2**     Types of Hydrostatic Bearings

After treating the principles on the example of the slipper, we shall now introduce two different classifications of hydrostatic bearings, the first one based on its geometric form, the second one on the load factor.

Hydrostatic bearings can be classified according to their form as follows:

1. Axial bearings, also called thrust bearings

2. Radial or journal bearings

3. Other bearing forms

The following observations can be made:

1. Axial bearings have plane bearing surfaces and are used to transmit axial forces. If misalignment or tilting of the bearing elements is avoided, the gap height is constant.

2. The surfaces or gap walls of radial bearings are built by cylinders with almost equal diameter. The gap is wrapped around and its height variable depending on the position of the journal within the diametric clearance.

3. Other forms have spherical or conical bearing surfaces, more difficult to manufacture and less frequently used. Such forms have the advantage that they can transmit radial and axial forces at the same time, /FN[1]/.

---

FN[1] — Amusing remark from the lecture : With this classification, one is at least sure not to overlook any type of hydrostatic bearings.

According to the load factor, the classification of hydrostatic bearings is as follows:

1.  Overbalanced bearings
2.  Underbalanced bearings

Here, the following should be noted:

1.  The load factor in overbalanced bearings is smaller than one ($\lambda < 1$) and the entire load, including any perturbing forces, is carried by the oil film as, for instance, in the case of the slipper in fig 1.

2.  The oil film in underbalanced bearings cannot carry the entire load and the rest is transmitted by metallic contact between the lips and the support, /FN²/. The depression contains the supply pressure and there is no need of a separate throttle. Nevertheless, a minimum impedance in the supply canal is desirable in order to prevent a complete separation on large disturbances. Underbalanced bearings will be treated in section 25−8.   Here, it should be mentioned that they can be characterised by a load factor of more than one.

It is possible to use an annular (i.e. in the form of a ring) or closed groove as represented in fig 2 instead of the depression both with overbalanced and with underbalanced bearings. The pressure development on the isle is similar to a depression but its surface must be included for the calculation of the tangential force by equation 1. Therefore, more heat is developed with high sliding velocities. Naturally, the groove must not run through the outer edges of the sealing lips to avoid a short circuit of the applied pressure.

The use of a closed groove instead of a depression according to fig 2 has the advantage that the surface for metallic contact is increased by the area of the isle, especially important with underbalanced bearings.

Real hydrostatic bearings have many different design forms but they are always derived from the basic form of a slipper.

---

FN² — A part of the remaining force can also be transmitted by hydrodynamical pressure with inclined slipper and sufficient sliding velocity, as described in 26.

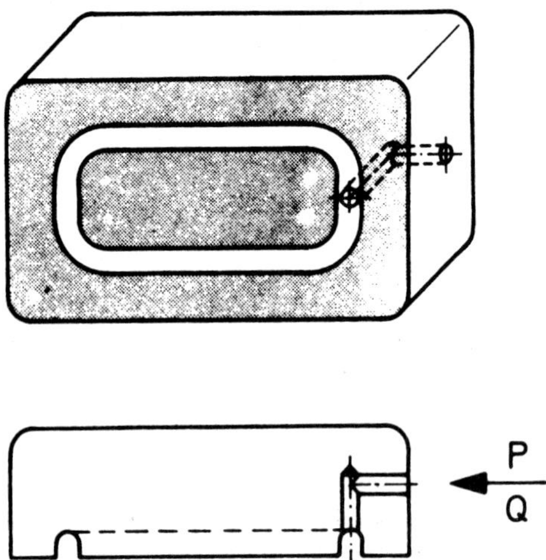

**Fig. 25.2: Slipper with isle in the depression**


### 25-3       Interaction of Variables

In order to gain a better understanding of the interaction of the operating variables in slippers and other hydrostatic bearings, we use the functional diagrams represented in figs 3 and 4. Fig 3 uses the supply pressure $p_{sp}$ as input variable as this corresponds to our intuition of the operation of a hydrostatic bearing and it contains the following operations:

1.  The applied pressure $p_o$ is determined by subtracting the pressure drop $\Delta p$ of the throttle from the supply pressure.

2.  The applied pressure produces the lift force by multiplication with the area $A_{tot} = A_{cv} + A_{lp}/2$ according to equation 1, as represented by the top block.

3.  The difference between the lift and the load force $F_{ld}$ gives an excess force.

4.  The excess force lifts the slipper and increases the gap height as represented by the block at the right with an interrogation mark since it is not very well defined. It depends on the compliance or softness of the suspension and also on accelerating

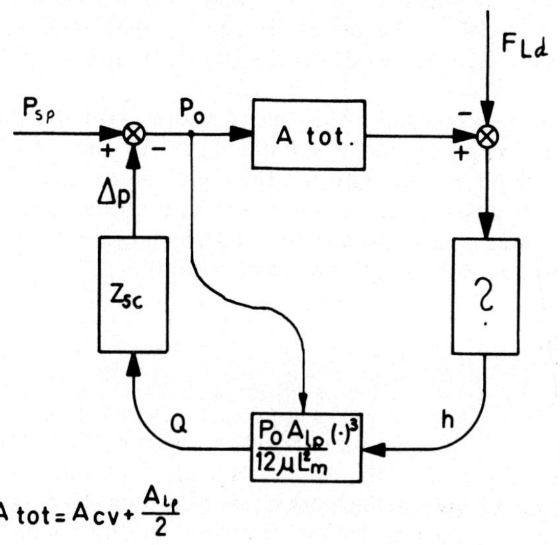



$$A_{tot} = A_{cv} + \frac{A_{lp}}{2}$$

Fig. 25.3: Functional diagram for the operation of hydrostatic bearings

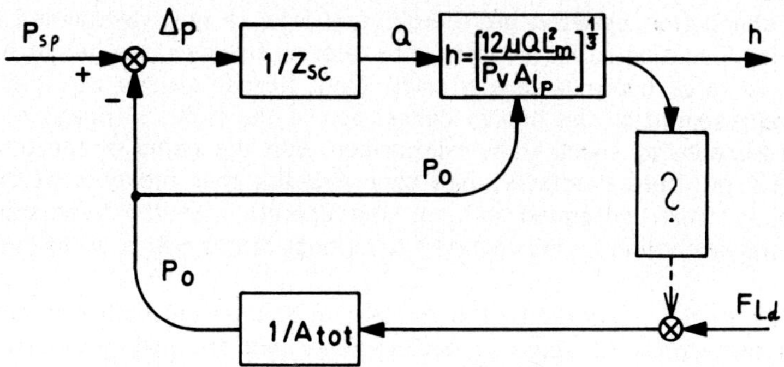



Fig. 25.4: Inverted functional diagram for calculation of hydrostatic bearings

forces. The compliance is usually large in the range of admissible gap heights and the acceleration forces small compared to the hydrostatic forces. Therefore, a small excess force gives a large change of the gap height leading to a large but not well determined gain in the block at the right.

5. The gap height determines the flow together with the applied pressure as shown by the lower block and by the cross connection. In order to calculate the flow, we use the gap formula applied to all sealing lips and introducing the mean gap length $L_m$ and the area Alp of all the lips surrounding the depression (i.e. at the right and left, in front and behind)

$$b = \frac{Alp}{L_m} \qquad\qquad Q = \frac{p_o}{12\mu} \frac{Alp}{L^2_m} h^3 \qquad\qquad (25-5)$$

6. The block at the left gives the pressure drop in the throttle as function of the flow. We use the secant impedance for the determination of general interaction and the tangent impedance for the study of small deviations from the equilibrium position.

Since fig 3 contains a closed loop with a high but not well known gain, it is convenient to invert it for further calculation, /FN³/, leading to fig 4.

The functional diagram in fig 4 is inverted and, therefore, the gains of the corresponding linear blocks are reciprocal and the relations in the non-linear blocks must be inverted. In particular, the connection between gap height and flow is now determined by solving equation 5. Furthermore, the relation between gap height and excess force becomes now a small gain, almost always negligible, as represented by the broken connection at the right. In practice, it is advisable to check it by comparison with the gains of the other blocks or, more precisely, one should check that the gain of this loop is small compared to one. The operation of the hydrostatic bearing according to the inverted functional diagram 4 is as follows:

The load $F_{ld}$ entered by the connection at the right bottom determines the applied pressure $p_o$ and, together with the supply pressure, the flow and, furthermore, the gap height. The reaction of the gap

FN³ — As explained in appendix 2.

height to the applied pressure is to be neglected as discussed. We have now a functional diagram especially suitable for computation where the different numerical values can be written below or above the connections.

A notable feature of the interaction of the variables is the fact that the load force is proportional — independently of the gap height — to the applied pressure $p_o$. This is a consequence of the linear pressure drop across the sealing lips, giving an average pressure of $p_o/2$ by equation 22–8, as long as the walls of the sealing lips are parallel. The gain between applied pressure and load force is then constant and denoted by $A_{tot}$ in figs 3 and 4.

With variable gap height, produced by misalignment of the slider or by not perfectly flat surfaces of the sealing lips, the form of the pressure distribution depends on the distance of the slider from the support. Whilst the effect of tilting will be treated in section 25–5, an interesting case is the linear decrease of the gap heights of a slider as can be obtained by warping of the surfaces of the sealing lips. The mean pressure over the sealing lips, as determined easily by one more integration of the pressure field in equation 23–4, depends on the minimum gap height. The detailed calculation and mean pressure curve will not be reproduced here, but the following limiting cases are of interest:

a) If the variation of the gap height along the leakage path is small compared to the minimum height at the outflow edge, the pressure field will be little affected and the mean pressure equal to $p_o/2$ as before.

b) If the gap height is reduced practically to zero and the slider touches its support with the outflow edges, the applied pressure will extend across the lips and the pressure drop concentrate at the contact line of the outflow edges and the support. The mean pressure is then approximately equal to $p_o$.

A slider with linearly decreasing gap height has therefore a stability in the pressure field without an impedance in the supply line since the mean pressure over the sealing lips can vary between the above limits.

Hydrostatic overbalanced slippers and bearings can effectively be built on this principle, but the difficulty is the manufacture of the variation of gap height which must be about equal to the minimum gap height, ie around ten micrometers. Practically the same effect is obtained if, instead of a linear reduction, a step is machined, for instance by photoetching, on the upstream half of the sealing lips such that the gap height is there the double of the height of the second part. Nevertheless, such small steps (about 10 micrometers) are easily worn away.

An increase of the gap height along the leakage path, on the other hand, leads to an opposed variation of the mean pressure and to instability of the slippers since the mean pressure becomes less with reducing distance from the support. We have, therefore, instability by the pressure field itself and the overall stability must be obtained by the variation of applied pressure with the gap height over the leakage and the impedance as represented by the lower part of fig 3.

Returning to gaps with parallel walls, the following two concepts of the stability of hydrostatic bearings should be distinguished:

1. Small perturbation stability.

2. Stability under large load variation.

The small perturbation stability is determined from the slope of the curve of the gap height as function of applied pressure (proportional to the load) and will not be pursued here.

Stability under large variations can be defined as the increase of the force produced by the pressure field if the sealing lips close up completely. In this case, the pressure in the depression rises to the supply pressure giving an increase of the force by $1/\lambda$, the reciprocal of the load factor (assuming that the form of the trapezoid pressure distribution is not changed). The large variation stability, is, therefore, proportional to $1/\lambda - 1$. If only a part of the depressions are fed over throttles and the rest without impedance in the supply, the stability factor is correspondingly reduced. Examples are fig 6 below and the valve faces of axial piston machines, treated in section 33-3.

Hydrostatic bearings supplied from a flow source and hydrodynamic bearings (section 26) produce in principle infinite forces on closing up and have, therefore, an infinite stability factor for large variations.

### 25-4    Circular Slippers

Many hydrostatic machines use circular or round slippers with a gap in the form of a disc between the sealing lips and the support.

As represented in fig 5, there is a depression in the centre with the radius $r_i$ followed by the sealing lip until the external radius $r_e$. We use here polar co-ordinates since the slipper is round or rotationally symmetric around its axis. It is convenient to consider the flow Q as given and to determine the pressure developed from equation 23–1.

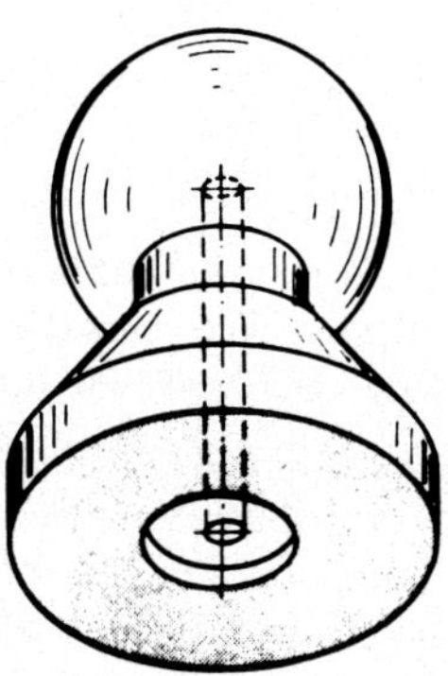

**Fig. 25.5: Perspective representation of a round slipper**

Instead of the variation of the gap height along the pressure gradient, as in section 23, we have here a variable width, but equation 23–1 remains valid. Inserting for the width of the gap the circumference $2\pi r$, we obtain

$$\frac{dp}{dr} = -\frac{6}{\pi}\frac{\mu Q}{r h^3} \qquad (25\text{–}6)$$

The differential equation 6 is separable and can be immediately integrated

$$p = -\frac{6}{\pi}\frac{\mu Q}{h^3}\left[\log r + \text{const}\right]$$

For the determination of the constant of integration, we note that on the external radius ( $r = r_e$ ) the pressure is zero leading to

$$p\,(r) = \frac{6}{\pi}\frac{\mu Q}{h^3}\log\frac{r_e}{r} \qquad (25\text{–}7)$$

Equation 7 has the property that the pressure becomes infinitely large at the zero radius or, in other words, there would be no flow with a finite applied pressure. In order to obtain a flow, it is necessary to limit the sealing lip at an internal radius $r_i$ and not to extend it to the axis of the slipper. The limit in figure 5 is the depression but, alternatively, the oil supply duct can limit the sealing lip.

The pressure in the depression becomes

$$p\,(r_i) \;=\; p_o \;=\; \frac{6}{\pi}\;\frac{\mu Q}{h^3}\;\log\frac{r_e}{r_i} \tag{25−8}$$

Equation 8 gives a linear relation between the applied pressure and the flow of a round slipper. The impedance depends only on the ratio of internal and external radii and not singly on these variables.

As we have seen before, the movement of the slipper has no influence on the pressure distribution with constant gap height or correct alignment to its support. On the other hand, the pressure distribution over the sealing lips produces a restoring moment if the slipper is misaligned, as shown by fig 7, independently of any sliding movement. In the case of a sliding movement, a further restoring moment is produced by hydrodynamical action. The alignment of the slipper can be disturbed by the friction forces or, rather, by the corresponding moment obtained by multiplication with the suspension height.

It is interesting to examine the pressure distribution according to equation 7 if the sealing lip becomes very thin or $r_e$ similar to $r_i$. We introduce the mean radius $r_m$ and $\Delta r$

$$r_e \;=\; r_m \;+\; \Delta r \qquad\qquad r_i \;=\; r_m \;-\; \Delta r$$

where $\Delta r$ is half the radial extension of the sealing lips and obtain

$$p_o \;=\; \frac{6}{\pi}\;\frac{\mu Q}{h^3}\;\log\frac{1 + \Delta r/r_m}{1 - \Delta r/r_m}$$

Thin sealing lips are characterised by $\Delta r/r_m \ll 1$ (very small) allowing us to expand the logarithm to terms of the first order

$$p_o \;=\; \frac{6}{\pi}\;\frac{\mu Q}{h^3}\;\frac{2\,\Delta r}{r_m} \;=\; 12\mu Q\;\frac{2\,\Delta r}{2\pi r_m} \tag{25−9}$$

Equation 9 is another form of our gap formula 22–4 using the circumference of the lips at their mean radius as width of the gap and their radial extension as length. As we shall see from the following numerical example, the approximate equation 9 is sufficiently accurate in all practical cases.

### Numerical Example of Circular Slipper

With $r_i = 1$ cm, $r_e = 3$ cm, an applied pressure of 200 bar, a gap height of $10\mu m$ and a viscosity of $\mu = 2 \cdot 10^{-7}$ barsec, we obtain the flow by inversion of equation 8

$$Q = \frac{\pi}{6}\ \frac{h^3}{\mu}\ \frac{p_o}{\log r_e/r_i} = \frac{\pi \cdot 10^9\ cm^3\ 200\ bar}{6.2 \cdot 10^{-7}\ bar\ sec}\ \frac{1}{\log 3}$$

$$Q = \frac{0.52}{\log^3}\ \frac{cm^3}{sec} = 0.47\ \frac{cm^3}{sec}$$

Using the approximate equation 9 instead, we have

$$2\Delta r = r_e - r_i = 2\ cm, \qquad r_m = 2\ cm, \frac{2\Delta r}{r_m} = 1$$

and

$$Q = \frac{\pi}{6}\ \frac{h^3}{\mu}\ p_o\frac{2\Delta r}{r_m} = 0.52\ \frac{cm^3}{sec}$$

The approximate formula 9 gives, with these dimensions, $r_e/r_i = 3$, a flow that is only 9% too large.

As a practical rule, even if the external radius is three times the internal, an error of less than 10% is committed by calculating with the approximate formula 9. This is the simple gap formula with the mean circumference of the sealing lips as width.

## 25–5    Tilting Forces and Tilting Moments

We have supposed until now that the gap height is constant across the sealing lips as can be attained by self alignment of the slipper or, with more difficulty, by accurate manufacture. The spherical head of the slipper on our figure allows self alignment, but we have to examine the question of the tilting stability as can be provided by the oil film.

Tilting moments can come from the following causes:

1. The tangential forces (equation 2) multiplied with the height of the centre of the sphere over the support.

2. Due to friction in the spherical joint if the support is slightly inclined or even changing its inclination, called also wobbling. It is produced by a disc not running square on a rotating shaft.

3. By eccentric load forces on slippers without spherical joint.

There are several methods to take up tilting moments without metallic contact:

1. With several depressions, each fed over a separate throttle as represented on the example in fig 6. The slipper has a central depression supplied with oil without throttle and two lateral depressions supplied over a throttle in the form of an adjustable needle valve. The pressure development is shown at the lower part in fig 6, for small gap heights in broken point, for medium gap heights in continuous and for large gap heights in broken lines. The pressure in the central depression without throttle is always equal to the supply pressure but it is variable in both lateral fields depending on the local gap height. With a small tilt or misalignment of the slipper, the gap height increases in one lateral depression and decreases in the other one. Therefore, the pressure field approaches on one side the broken point and on the other side the broken line in fig 6. This produces a restoring moment, the magnitude of which can be easily estimated from the areas of the depressions.

2. The effect according to 1 above is produced especially if one uses a larger bearing surface with many depressions instead of several slippers, each fed over a separate throttle. In the case of a flow source, it is necessary to use a multiple pump, i.e. a pump capable of supplying several flows with different pressures, because any parallel connection between the depressions would destroy the restoring moment.

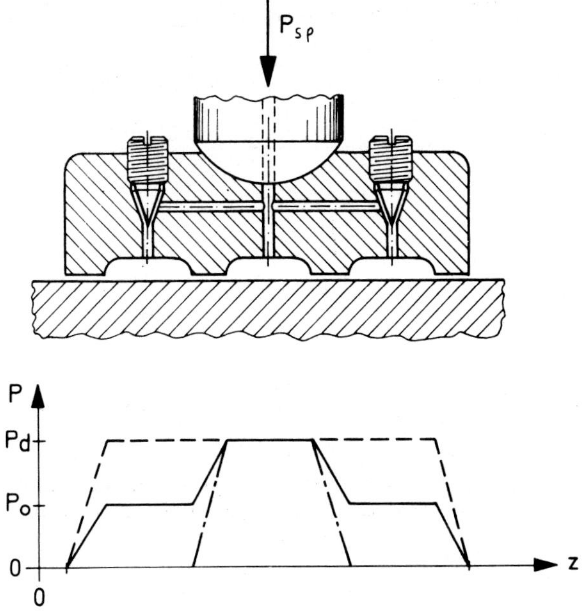



Fig. 25.6: Cross-section and pressure development in
a slipper with several depressions for stabilising
against tilting moments

3. If the sealing lips have sufficient radial extension in the direction of the pressure gradient, one has an asymmetric pressure field on misalignment, as treated in section 23. As shown
   in fig 7, it is convex at the smaller gap height since the
   pressure gradient is concentrated at the downstream edges. On
   the lip with the greater gap height, there is a concave curve
   with the pressure drop concentrated at the upstream edges. We
   see easily how these deviations from the linear pressure drop
   across the sealing lips give a restoring moment.

We can estimate the restoring moment by assuming that the pressure
distribution is undisturbed at the lip with the greater gap height and that
the applied pressure extends until the downstream edge at the lip with the
lower gap height. It gives a restoring moment equal to the product of one
half of the lip area, the applied pressure and the two thirds of the radius
of the lip. Details of this calculation are beyond the scope of this book.

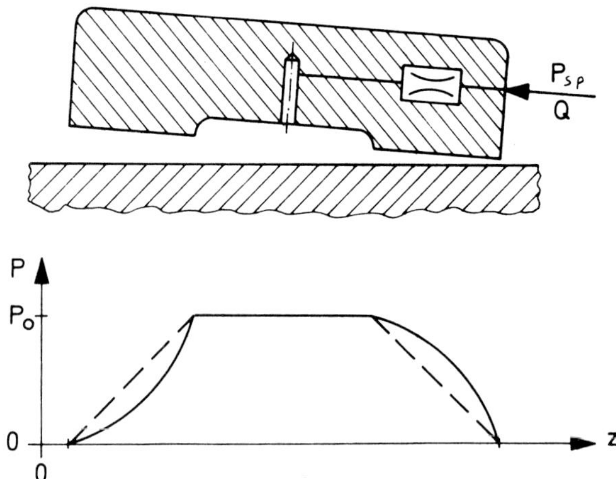

Fig. 25.7: Stabilisation of a slipper against tilting
moments by the pressure distribution on extended
sealing lips

## 25–6      Geometric Design and Load Capacity

The main condition for the design of hydrostatic bearings is reliable operation without seizing or wear. Furthermore, the losses or the power consumption of the bearing should be as small as possible. The security of underbalanced bearings depends essentially on the choice or combination of materials. From a design point of view, it is advisable to avoid edge pressures or local spots of high contact forces and to transmit the forces as continuously as possible over the contact which should be as large as possible.

The most important condition in overbalanced bearings is to maintain a certain minimum gap height. The sliding surfaces are then always separated by an oil film but the choice of material still has its importance for taking up any metallic contact during exceptional conditions. In normal operation, the security of the bearing increases with the gap height.

In principle, the desired gap height can be produced with any load as long as the supply pressure is sufficiently high. In other words, the load factor as defined in equation 1 must not come too near to one.

We shall now calculate the power consumption including the losses in the throttle. Inserting the power due to the tangential force from equation 2 and the power due to the flow from equation 4, we obtain

$$P = \frac{\lambda p^2 sp}{12\mu} \; \frac{b}{L_m} \; h^3 + \mu \, b \, L_m \; \frac{v_0^2}{h} \qquad (25-10)$$

We see in equation 10 that the first term increases and the second term decreases with increasing gap height. Therefore, there is a certain gap height where the power consumption becomes a minimum. Its calculation is simple but not reproduced here because of its small practical importance. The optimum is a function of the operating variables, especially of the sliding velocity, and of the supply pressure, usually variable in hydrostatic machines. Therefore, it is not possible to design or optimise the bearing in such a way that the gap height with minimum losses is maintained.

The design of hydrostatic bearings aims at obtaining as much gap height as possible with a given load and flow. According to formula 10, the width or extension in the circumferential direction b should be small and the length $L_m$ in the direction of the pressure gradient large.

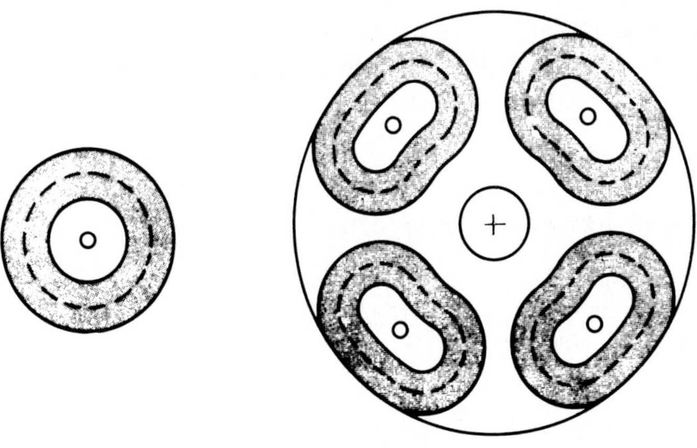

**Fig. 25.8: Bottom view of well designed fully developed sealing lips**

Fig 8 contains two design examples according to these recommend-
ations, at the left the bottom view of a slider and at the right, half
of a hydrostatic bearing with four oval supporting fields. The sealing
lips are in both cases well extended' in the direction of the pressure
gradient and the mean circumference, as drawn in by the broken line,
is as small as possible. The mean circumference is situated approximately on the middle of the sealing lips and we can easily draw it
in a given design by inspection.

The mean circumference, drawn as broken line in fig 8, limits the
area to which pressure is applied. Its length is equal to the width in
the gap formula 22−4. The circumference should be as short as
possible with a given area or, as known from geometry, approximate
to a circle.

Whilst it is useful in principle to make also the length of the
lips, ie their extension in the direction of the pressure gradient,
rather large, this is limited in practice as follows:

1.  The entire force transmitted is composed of the pressure fields
    over the depressions and over the sealing lips according to
    equation 1. However, the pressure field over the lips is subject
    to the mentioned perturbations of the linear pressure drop. As
    we have seen in section 22, the pressure development can be
    disturbed, especially with minimum tilting or misalignment of
    the bearing part, or the factor 1/2 in equation 1 may change
    correspondingly. It can vary between the limits of zero (entire
    pressure drop at upstream edge) and one (entire pressure drop
    at downstream edge) and Guillon, B 7, denotes the lips as a
    doubtful area. Because of these doubts, it is not advisable to
    give to the lips too large a share, for instance, more than 50%
    of the forces, especially with high load factors. Geometrically
    speaking, the lip and depression area should be smaller than
    the load area $(A_{lp} + A_{cv} < A_p)$ limiting the gap length with a
    given circumference.

2.  With increasing length, the temperature produced within the
    lips increases also (equation 24−4). This is dangerous, as
    will be seen during the discussion about the necessary gap
    height below. One remedy is, with hydrostatic bearings with
    several depressions, not to let the sealing lips run through but
    rather to apply relief or rinsing grooves, as shown in fig 8.

Concluding about hydrostatic bearings, we should like to mention
that they have many problems in common with hovercraft vehicles.

There also it is important to design the supporting area with its skirts in such a way as to obtain as much clearance over the ground as possible with a given air flow. The clearance must be so large that the irregularities of the supporting surface, for instance waves on water, cannot touch the vehicle or the skirts in order to avoid damage. Furthermore, it is necessary to design sufficient stability against tilting which could lead to touching of the ground with consequent damage.

Common advantages of hovercrafts and hydrostatic bearings are small friction forces, especially with small sliding velocity. The power consumption due to feeding with lubricant is an accepted disadvantage.

### 25—7    Necessary Gap Height

We have seen that increasing the gap height increases the security but also the power consumption of the flow according to equation 22—6. Therefore, we should like to use as little gap height as possible depending on the following considerations closely connected with the practically admissible load:

1. The gap height must be larger than the roughness of the surface and the possible distortion of the support under load. From this it becomes important to make the supporting parts stiff and to limit roughness. Objectionable deformations include thermal expansion, especially of pistons in cylinders. Such thermal expansion can easily become comparable with the gap height, especially in large machines.

2. The gap height should be larger, ideally twice as large, than the largest impurities in the lubricant. Otherwise, one obtains the so-called abrasive wear and, with much contamination, even seizing of the lips. It is generally accepted that impurities a little larger than the gap height are the most damaging because they squeeze between the lips. Much larger impurities cannot enter at all and smaller impurities are less damaging, but they contribute to erosion. It follows therefore that smaller gap heights can be admitted with improved filtration.

3. The heat produced increases with smaller gap height, whilst the heat carried away by the leakage decreases. If there is no heat conduction through the wall, the temperature increase of the oil is determined by equation 24—4 and, for safe running, it

is advisable to limit it to 20–30° C. We see therefore that, with increasing sliding velocity and gap length, a larger gap height should be chosen.

All these considerations impose a gap height of the same order of magnitude, mainly 8–10$\mu$m with smaller and 13–15$\mu$m with larger hydrostatic machines. Nevertheless, there exist several hydrostatic machines and components with a very small gap height — in the range of 1–5$\mu$m — but they must rely on a very good filtration of the fluid and of evacuation of the heat by the walls.

**25–8**    Design of Underbalanced Slippers and Bearings

Underbalanced hydrostatic bearings are, strictly speaking, only a kind of hydrostatic unloading device where the remaining forces are transmitted by metallic contact, as mentioned in section 25–1. They are used frequently in oil hydraulic practice despite their more uncertain performance and will be discussed briefly here.

It is convenient to introduce a *balancing factor* $\beta$. This factor is the fraction of the total load carried by the pressure field assuming the trapezoid pressure development as described in section 25–1. Since the pressure in the depression is equal to the supply pressure, we have as definition

$$\beta = p_{sp} \left[ A_{cv} + A_{lp}/2 \right] / F_{ld} \tag{25–11}$$

The balancing factor $\beta$ is, in a certain sense, the reciprocal of the load factor $\lambda$ introduced by equation 1. It is always less than one corresponding to a load factor greater than one. In fact, a balancing factor of more than one would imply an overbalanced bearing where the impedance in the supply duct is required in order to prevent separation. The balancing factor is strictly equal to the reciprocal of the load factor only if supply pressure and applied pressure are equal as it is the case in underbalanced bearings. Nevertheless, it is convenient to introduce this new symbol.

The remainder of the load forces not carried by the pressure field is transmitted by metallic contact of the sealing lips. This produces solid friction with the usual friction coefficient giving a tangential force independent of the sliding velocity, as a first approximation, similar to ordinary dry metallic friction.

It is interesting to calculate the tangential forces or the reduced coefficient of friction in underbalanced bearings. The viscous forces of the fluid in the depression are negligible compared to the metallic friction of the lips in contact transmitting the fraction $1 - \beta$ of the total load. This leads to

$$f_{rd} = f\hat{\mu}\,(1 - \beta) = 0.1\,(1 - 0.9) = 0.01 \qquad (25-12)$$

The third and fourth form of equation 12 were obtained by assuming $\beta = 0.9$ and $f\hat{\mu} = 0.1$ as representative values in hydrostatics.

The coefficient of metallic friction $f\mu$ is as doubtful and variable as with normal friction. According to Wüsthoff /FN$^4$/ one can count on friction coefficients between 0.05 and 0.1 whilst Vogelpohl /FN$^5$/ gives values of 0.14 to 0.17 for large machines. Both indications are given for the case that a large part of the force is carried by an oil pressure field. Therefore, the mean friction coefficient of 0.1 appears reasonable leading to a reduced friction coefficient of 1% with a balancing factor of 90%.

Whilst the reduced friction coefficient of 1% appears reasonable, it is still about ten times larger compared to what can be reached in overbalancing, as seen by comparison with equations 3 and 3a.

Due to the possible variations of the mean pressure, it is advisable to make the area of the sealing lips not too large. Ideally, it should be

$$Alp < 2\,(1 - \beta)\,\frac{Flp}{p_{sp}}$$

because this prevents separation even if the supply pressure extends to the outer edge. In the above example, this imposes that the sum of the areas of the lips and of the depression is smaller than the load force divided by the supply pressure. In order to increase the area for the metallic force transmission, additional contact pads can be provided.

Fig 9 shows an underbalanced slipper with sealing lips similar to the overbalanced slipper in fig 6. The inner continuous ring constitutes the thin sealing lip whilst the outer ring provides additional contact pads. It is broken by radial slots in order to remove any pressure in the circular groove and should be compared to fig 1.

FN$^4$ — P Wüsthoff, 'Hydrostatische Flügelpumpen', thesis TH Eindhoven 1969.

FN$^5$ — G. Vogelpohl, 'Betriebssichere Gleitlager', Springer, 1956.

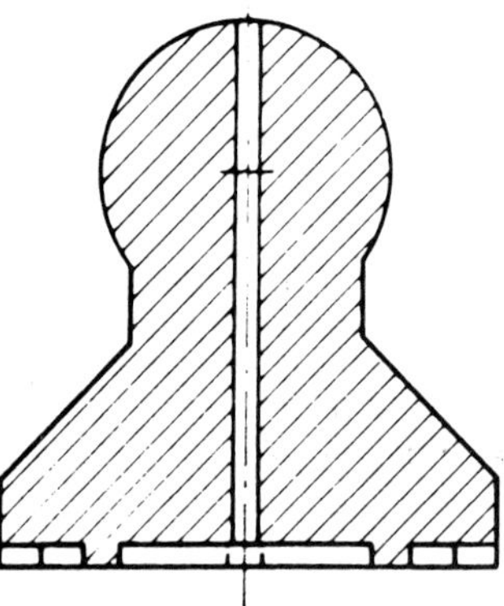

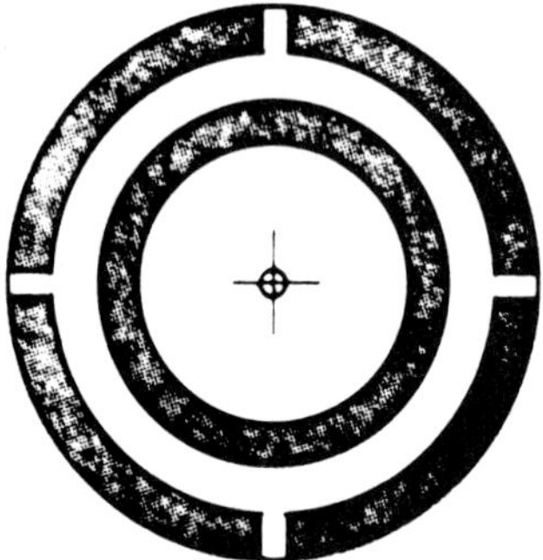

Fig. 25.9: Underbalanced slipper with inner sealing rings and with outer contact pads

Underbalanced slippers can give satisfactory results with not too high sliding velocities since the lubrication of the contact faces is very good. The reliability of the metallic contact depends essentially on the selection of materials. In hydrostatics, a combination of bronze and hardened steel is used — just as with overbalanced slippers — but wear and contamination sensivity remains a great question. According to Vogelpohl, the metallic contact always produces a measurable wear that can be tolerated only for short-life applications.

The sealing lip works here under boundary lubrication in order to reduce wear. It is advisable to retain some controlled roughness so that it may take some lubricant into the immediate proximity of the contact points. Furthermore, it is important that the lubricant wets the sliding surfaces due to its surface tension. The power balance, as treated in Section 24–1, is applicable also with boundary lubrication, but the power transport connected with the leakage or the oil flows $P_1$ and $P_4$ becomes negligible. In order to calculate the power transport due to shear, one has to use the tangential force given by the product of the normal force or load and the coefficient of friction.

In order to evacuate the power under boundary lubrication, only the heat conduction of the walls is practically available. One will always try to assist it by design measures as, for instance, by suitable cooling ducts with a flow of fresh oil. Another method is to provide that each spot of the surfaces is loaded only intermittently and acting as sliding surface and unloaded and rinsed with fresh oil in between. This is the case with the swashplate of axial piston machines (see Section 33–1) but not with the sealing lips of the slippers.

## 26   Hydrodynamic Bearings

### 26—1   Geometric Wedges

Hydrodynamic bearings comprise essentially a gap with decreasing height, producing normal forces due to the pressure field without metallic contact. We have studied the pressure development already in Section 23—2 and shall apply this now to the bearing represented in fig 1.

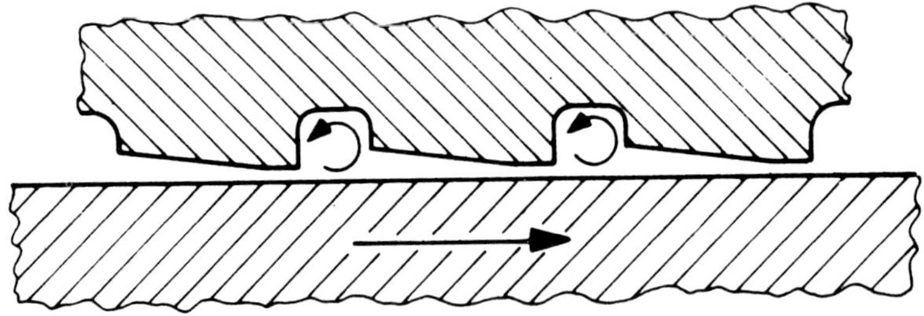

Fig. 26.1: Cross-section through hydrodynamic, inclined supporting pads

Fig 1 contains a number of pads with slightly inclined contact faces. They are not extended across the entire contact surface but leave a trailing part without inclination as shown. This avoids very high contact pressure in exceptional cases when no oil pressure is developed as, for instance, when starting up. Otherwise, we would have a line contact or cutting in of the trailing edge into the support. We shall neglect the influence of the not inclined part for the calculation. It should also be noted that pressure can be developed only in one direction of movement or rotation. We use equation 23—9 for the calculation and insert there the linear gap height according to equation 23—3. Then, equation 23—9 can be considered as a differential equation and the oil flow determined from the boundary conditions for the pressure. They state that the pressure at the leading and at the trailing edge of the pad are zero.

Another possible method would be to consider the flow as given and to calculate the pressure development. For inversion, we would select such a flow that the pressure at the leading edge of the pad becomes zero.

We obtain a pressure distribution by solving equation 23–9 with a maximum slightly behind the centre of the pad (see Bibliography 9, chapter 12, Section 5–4) as function of the minimum gap height at the trailing edge $h_o$ and the gap height at the leading edge $h_1$ .

We are mainly interested in the forces exerted on the walls obtained by integration of the pressure field. We can represent the result as follows

$$F_{ld} = p_{ld} \, b \, L \qquad\qquad p_{ld} = \frac{\mu \, v_o \, L}{h_o^2} \; f \left( \frac{h_o}{h_1} \right) \qquad (26\text{–}1)$$

Here, $p_{ld}$ is the mean fluid in the pressure bearing depending on the product expression and on the function of the ratio $h_1 / h_o$ represented in fig 2.

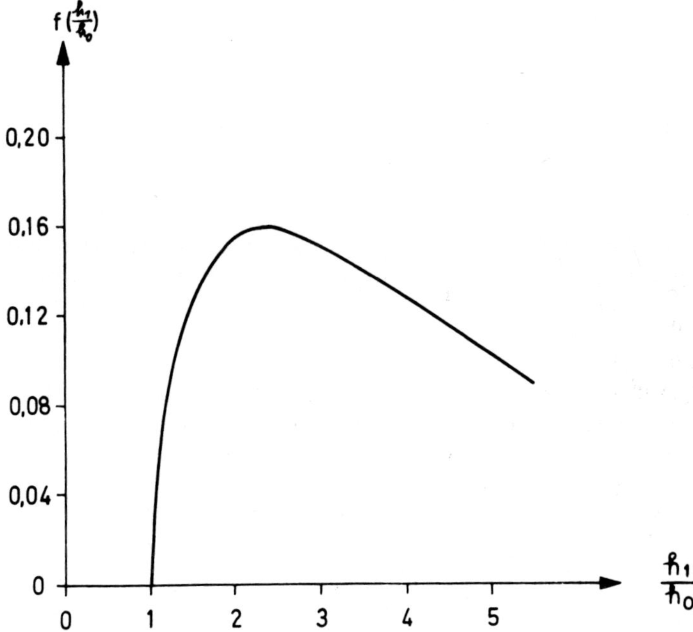


**Fig. 26.2: Load forces of the hydrodynamic bearing according to equation 26-1**

**Numerical Example of Hydrodynamic Bearing**

What bearing pressure is produced with a pad length of 1 cm, a minimum gap height of $h = 10$ micrometer $= 10^{-3}$ cm with a sliding velocity of 500 cm/sec?

We see from fig.2 that the highest load or bearing pressure is carried with a height ratio of $h_1/h_o = 2 \cdots 3$, the function becoming then $f = 0.16$. The bearing pressure becomes

$$p = 0.16 \; \frac{\mu v_o L}{h_o^2} \qquad\qquad (26\text{–}1a)$$

$$p = 0.16 \cdot 2 \cdot 10^{-7} \text{ barsec } 500 \; \frac{\text{cm}}{\text{sec}} \cdot \frac{1 \text{ cm}}{10^{-6} \text{ cm}^2} = 16 \text{ bar}$$

We obtain a bearing pressure of 16 bar with a wall inclination of

$$1 \cdots 2 \; \frac{10^{-3} \text{cm}}{1 \text{ cm}} = 1 \cdots 2 \text{ milliradiants}$$

For a vivid picture of the operation of hydrodynamic bearings, we consider the pads loaded by a constant force and maintained in the required inclination $h_1/h_o = 2 \cdots 3$.

The pads will lift off until formula 1 is satisfied. The gap height increases with the velocity whilst, without velocity or with stopped bearing, no lifting takes place. The admissible load increases thus with the speed. Hydrodynamic bearings operate similarly as water skis where the lift is produced by the speed and by slight inclination of the skis. The difference is that the lift on the water depends mainly on the acceleration forces and is connected with the free surface of the lake.

The pressure development as described is somtimes called a geometric wedge since the oil film acts a little like a wedge. The admissible load of a given bearing depends strongly on the minimum gap height to be selected according to Section 25–10.

It is useful to remember always the assumptions involved in setting up equation 1 :

1.  The viscosity was assumed constant while, in reality, the oil is heating up due to the tangential forces. This can be introduced approximately by using an effective viscosity, ie the viscosity of the oil at the mean temperature between entrance and exit.

2.  The supporting pads had been assumed infinitely large whilst they are sometimes only a little broader than they are long or even square. Therefore, we have an outflow sidewise reducing the pressure field. Its exact calculation is rather difficult, but we note here only that, with square pads, the side flow reduces the lift force by about 50%. More exact indications are contained in Section 5—8 of Bibliography 6.

3.  We have assumed the walls to be perfectly rigid whilst deformations begin to become sensible with very high local pressures.

The depressions between the pads are very important since they determine the pressure at the leading edges of the pads and feed there cooled and fresh oil.

The usual terminology for this lift generation, hydrodynamic bearing action, is rather unfortunate. As we have seen, the pressure development is due to the viscosity forces which draw the oil into the progressively decreasing gap height and the word 'dynamic' should be used only for inertia forces. The only consolation is that the accepted word 'thermodynamics' suffers from the same defect since it is concerned with equilibrium states of thermal systems.


## 26—2    Different Design Forms


The basic arrangement of fig 1 can be considered as the development of an axial bearing. The main difficulty in manufacture is the very small inclination of the pads, but some specialised manufacturers produce such hydrodynamic bearings in standard sizes, similar to roller bearings.

Hydrodynamic radial bearings are more practicable since the progressively decreasing gap height is formed by two cylinders of almost equal diameter. The cylinders are constituted by the shaft and the journal and the diametric clearance determines the gap height together with the actual position of the shaft in the bore. Unfortunately, we cannot treat these interesting questions in more detail.

The so-called thermal wedge is another form of hydrodynamic bearing. Here, the pressure is produced by the fact that the oil temperature increases whilst flowing through the gap and the consequent thermal expansion produces a pressure field, similar to the geometric wedge. A comprehensive calculation is very difficult because the compressibility of the oil and the temperature dependence of the viscosity must be included, but it is possible to obtain gap heights of several micrometers. /FN¹/

We would like to note here that the thermal wedge is based on the temperature increase of the oil within the gap and that, with high and constant temperature, no pressure field is produced. Therefore, the depressions between the pads with cool, fresh oil are especially important here.

The exact form of the gap is of little importance as has been calculated by Rayleigh (see Bibliography 4, Section 5-3). With given minimum gap height, practically the same lifting force is obtained with other forms of height reduction, for instance, if the pads have a step producing over the first half of their length twice the gap height of the second half.

Instead of fixed inclination, it is possible to suspend the pad over a joint, similar to fig 25-1 and to rely on self-alignment to produce the required inclination. This principle has been introduced by Mitchell in the bearing carrying his name since the beginning of this century. These bearings are used especially for the propeller thrust of large ships. Mitchell found out that the centre of gravity of the pressure field migrates to the rear with increased inclination producing correct self-alignment. /FN²/.

As with other hydrodynamic bearings, the Mitchell bearing can carry no load with very small rotation frequencies or sliding velocities, for instance when starting up. This is disturbing, especially for bearings with loads independent of rotation frequency such as the weight of the rotor in water turbines. In contrast, the thrust of ship propellers is proportional to the square of the rotation frequency and insignificant at starting so that it is well suitable for a Mitchell bearing. In some large installations, even with hydrodynamic main bearings, one provides a hydrostatic ancillary lubrication by a small

---

FN¹ — The reader is referred to Bibliography 4, Section 1-3-8, for an approximate, and to Bibliography 1, Section 4-3, for a complete treatment.

FN² — For the not difficult calculation refer to Bibliography 9.

pump switched off after starting. This is called jacking and used frequently with large steam turbines. /FN³/

For completeness, it should be mentioned that a pressure field is also developed if the gap walls close upon each other because the oil trapped between them must be expelled. The simple theory of this effect, also called squeeze film, is not reproduced here because of its insufficient importance in hydrostatic machines where it usually contributes only to damping of possible oscillations of bearing members. On the other hand, the shock loads produced by gas pressure in internal combustion machines are largely taken up by squeeze films.

### 26-3    Lubricating Wedges in Roller Bearings

A cylinder rolling on a support or two cylinders rolling between themselves as, for instance, in roller bearings and in gear, constitute a gap with progressively decreasing height. Therefore, we can expect there also the production of a pressure field.

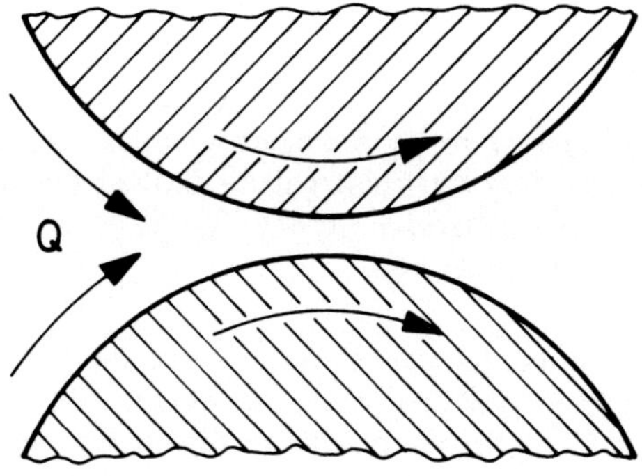


Fig. 26.3: Cross-section through rigid rollers forming
a small oil film

FN³ — For further information on turbine bearings, see D.M.Smith, 'Journal Bearings in Turbo-Machinery', Chapman and Hall, 1966.

Fig 3 represents the basic arrangement. In the region of minimum gap height, where we expect a pressure, only the difference of the radii is of importance and not the fact whether it is formed by a cylinder and a plane or by two cylinders. For the computation of the pressure field, one has to take into account that both walls are moving here. The two moving parallel walls would entrain, without pressure, the following flow

$$Q = v_0 \, b \, h \qquad\qquad (26\text{–}2)$$

Compared to the similar equation 21–2, the factor 1/2 has disappeared because of the movement of the support. With variable gap height, we may add this flow to the flow due to the pressure gradient obtaining instead of equation 23-9

$$Q = v_0 \, b \, h - \frac{b\,h^3}{12\mu} \frac{d\,p}{d\,z} \qquad\qquad (26\text{–}3)$$

As just mentioned, the factor 1/2 has disappeared compared to equation 23–9 due to the movement of the support. Therefore, we obtain the same pressure field with one half the wall velocity and the same other variables. All solutions of equation 23–9 become thus solutions for equations 26–3 by the simple substitution $v_0 \to v_0/2$ . /FN[4]/

Our rolling cylinder or roller produces a pressure field by hydrodynamic bearing action at the most narrow part of the gap. After the contact point, we shall generally expect cavitation but, generally, no damage is done in roller bearings.

The hydrodynamic lubrication of rollers has the property that the force increases strongly with decreasing height, just as with fixed pads. Calculating then the gap height in this model under the usual loads in roller bearings, we obtain very small values, of the order of 0.02 micrometer, negligible compared to the surface roughness and to other deformations.

---

FN[4] — Equation 26–3 has been built here by the reasonable composition of two flows. It can be derived rigorously from the Navier-Stokes equations with the same assumptions as in Section 23 but with the boundary conditions $v(x = 0) = v(x = h) = v_0$

Newer researches have shown that it is not realistic to consider the walls as perfectly rigid in the contact zones of the rollers. Already without lubricant, the rollers are flattening in the contact zone according to the Hertz formulae producing a small plane contact area. The oil entering this area finds a gap with flat walls as in fig 4. The enlargement of the contact area favours the building of the pressure field and increases the gap height at the same load by almost two orders of magnitude — to about one micrometer. This interaction between wall deformation and pressure development is called elasto-hydrodynamic lubrication and treated in the excellent book of D. Dowson and G. Higginson, 'Elasto-hydrodynamic Lubrication', Pergamon Press, 1966. The computation of the interacting pressure, elastic stress and deformation fields is very difficult and can be done only with large computers. An interesting result is that the temperature increase in the oil film is very small, probably due to the small absolute dimensions and the continually changing contact areas, allowing calculation with constant viscosity. As a conclusion, cylinders in contact, like rollers in bearings and teeth of gears, are practically always separated by an oil film.

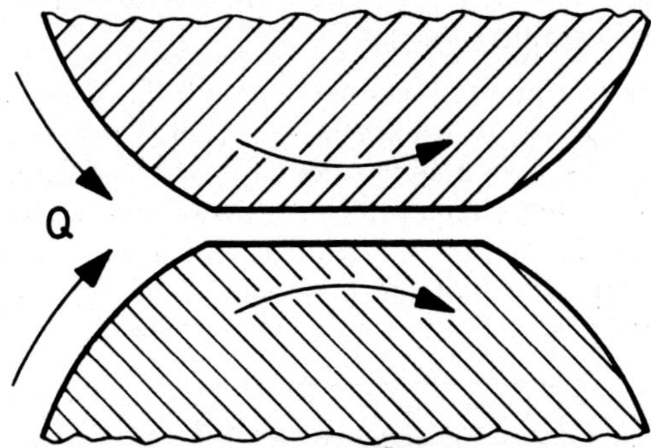


Fig 26.4:Cross-section through rollers flattened by contact pressure forming a more extended oil film

## 27  Stability of Pistons and Spool Valves

The pressure development in gaps is important also for the lateral
stability of spool valves and pistons in cylinders since the clearance
is subjected to high pressure gradients. Examples are the pistons of
hydrostatic machines and the spools of servovalves and other valves.
The moving parts have the tendency to stick to the cylinder walls,
also called hydraulic lock. This phenomenon can be avoided by a
number of annular grooves as we shall see.

We know that the pressure gradient in a gap with variable height
is not constant but increases with smaller gap height. The average
pressure gradient becomes just large enough to build the applied
pressure over the length of the gap.

For simplicity, we do not consider a round piston but rather a flat
slider arranged in a slit as shown in fig 1. High pressure is applied
from the left and we consider the possibility of upwards and downwards movements of the slider within the clearance, whilst any tilting
will be treated at the end of this section. We assume that the slider
is slightly conical such that the clearance at its left end, the high
pressure side, is less high.

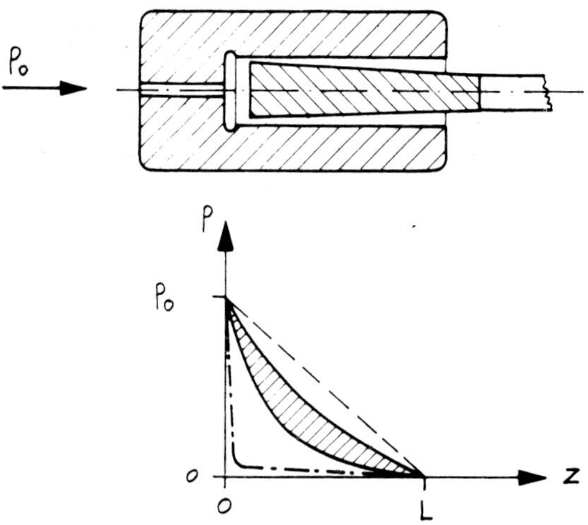



Fig. 27.1:  Cross-section and pressure development in
a conical slider with slightly increased thickness at
the high pressure side

The lower half of fig 1 shows the pressure development in both gaps. Assuming that the slider is displaced a little to the bottom from its central position, the lower gap is especially narrow at its left end and the pressure development follows the lower full line. In the upper gap, the ratio of the gap height at the entrance and the exit or left and right is less different from 1 and the pressure develops therefore according to the upper full line. The net force on the slider is equal to the hatched area between both pressure curves and presses the slider to the bottom whilst it would be pressed to the top with a small displacement above the centre. The slider is, therefore, unstable and has a tendency to stick to the walls. In the limiting case, with the left end of the slider touching the wall, the entire pressure drop is concentrated upon the contact point, as shown at the point-broken line. The pressure development in the upper gap approaches then the linear drop, as shown in fig 1 in broken lines. It is easy to compute the pressing force for this case since we have, on top, the force corresponding to linear drop according to equation 22–7 and, on bottom, no hydrostatic force

$$F = \tfrac{1}{2}\, p_0\, b\, L \tag{27-1}$$

with $L$ = length and $b$ = width of the slider. The normal force can thus grow until one half of the product of applied pressure and area of the slider. Multiplying by the reasonable friction coefficient of 0.1, the sticking force is obtained. It is very large compared to the actuating forces and can lead to failure of the slider or of a valve by sticking.

With the usual round pistons and sliders, we obtain similar conditions but, according to Guillon, Bibliography 7, appendix 4–2, the normal forces are somewhat smaller, at most 27% of the product of applied pressure, length and diameter of the piston.

If, on the other hand, the slider is somewhat more slender on the high pressure side – fig 2 – then we have the pressure development as drawn below. If the slider is displaced below from the centre, the pressure drop in the lower gap concentrates on the right hand side, as shown by the upper curve, and, in the upper gap, it is more evenly distributed as shown by the lower curve. Therefore, we obtain a net force to above, corresponding to the hatched area between both pressure curves, tending to return the slider in its central position.

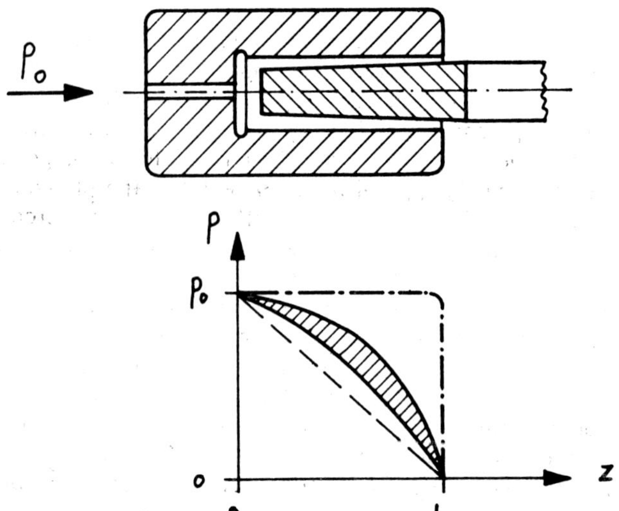

Fig. 27.2: Cross-section and pressure development through a slider with slightly increased thickness at the low pressure side

In order to remember the stability conditions, imagine that the cork of a wine bottle decreases its diameter at the internal end and the plug of a champagne bottle increases. According to our considerations, a plug for a wine bottle should be stable and a champagne plug should be unstable. It would stick to the walls — very useful for taking up the gas pressure of champagne.

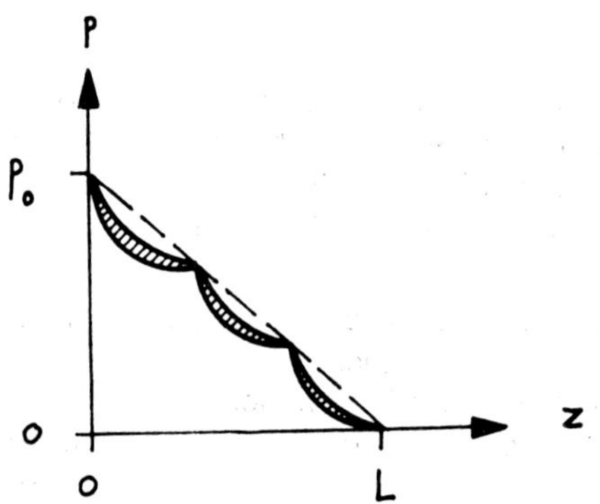

Fig. 27.3: Pressure development on sliders and pistons with annular groove or compensating bores

Due to manufacturing difficulties, conical pistons are not frequently used in practice. Instead, one uses annular grooves with pistons and spools and cross bores for flat sliders in order to equalise the pressure at both sides for stabilisation. Fig 3 shows the pressure development with annular grooves leaving only very small net forces. The cross-section of the grooves (or cross bores) must be large enough to take the flow without sensible pressure drop. In principle, lengths and depths of several hundreds of micrometers would be sufficient but, for ease of manufacture, they are usually made somewhat larger.

Sometimes, **one** uses annular grooves in cylinders instead of (or together with) grooves in pistons, especially with axial piston machines despite more difficult manufacture. This has the advantage that any impurities can settle down in the grooves of the cylinder under the centrifugal forces of the rotating cylinder block. They are not ejected as with the groove of the pistons where they would be drawn by the flow into the gap. It appears that annular grooves in the cylinders increase the impurity or contamination resistance of axial piston machines. For more simple manufacture, such grooves are sometimes made not as a complete annulus but rather around a part of the circumference in the form of a quarter-moon.

# 3 HYDROSTATIC PUMPS AND MOTORS

## 31    Operation and Operating Limits

### 31—1    Production of Hydrostatic Power

In this chapter we shall consider the production of hydrostatic power or of flow under pressure in general. One uses principally hydrostatic pumps, most of which can be used also as motors for reconversion of hydrostatic to mechanical power. Therefore, they are called more appropriately: hydrostatic machines.

The design of hydrostatic machines is governed by the following properties of the working fluid:

1. The fluid is practically incompressible, and expansion and compression have much less importance as for instance with air compressors.
2. The liquid is lubricating and clean, and can be used for the lubrication of the sliding surfaces /FN[1]/
3. The machine must be able to support high pressures without excessive deformations.
4. In many cases it is desirable to provide an adjustable displacement or volume delivered at each revolution or radiant of rotation on the shaft.

---

FN[1] — This is quite the opposite to four stroke internal combustion engines, where the working fluid and the lubricant must be carefully separated.

With high operating pressures, from 70 bar onward, and with small flows, usually not more than 10 lit/sec, turbo-machinery is not very suitable, and therefore only positive displacement machines are used.

Positive displacement machines have been known for a long time and work as follows : A displacement chamber, for instance, a cylinder with a piston assumes alternatively a large and a small volume. It is connected with the intake tube during volume increase for filling up, and switched to the delivery tube displacing the liquid on volume reduction. The necessary switching, in oil hydraulics also called timing or distribution, is done in the traditional water pump by non-return valves. Hydrostatic machines contain normally a distribution by ports or slits that register at the correct rotative position. This is called positive distribution and will be treated in detail in section 33–4.

Many basic mechanisms have been proposed as displacement machines since the time of steam engines (see for instance B 12). Today the important basic mechanisms for hydrostatic machines can be classified as follows:

1.  piston machines
2.  gear machines
3.  vane machines
4.  other machines

Whilst there are, in principle, many mechanisms in the class of 'other machines', only the so-called screw pumps have reached practical importance.

Although the working liquid is practically incompressible, we shall see when discussing the working diagram below that the remaining compressibility controls the pressure development during the change from admission to delivery port.

### 31–2   Working Diagrams

The displacement chamber of variable volume is found in all hydrostatic machines in more or less hidden form. We shall study at first the operation on the example of a simple piston in a cylinder, as represented on fig 1.

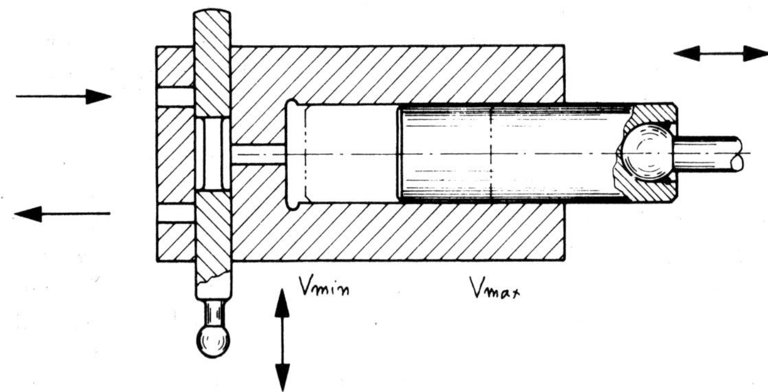



**Fig. 31.1:Cross-section through a cylinder with the dead centres of piston movement and with distribution by sliders (left)**

Fig 1 contains a piston in a cylinder connected by a slider in the centre with the intake port or tube (at left above) and with the delivery port at the left below. The dead centres of the piston movement are designated with $V_{min}$ and $V_{max}$, since they correspond to the extreme values of the displacement volume.

The working diagram or relation between volume and pressure in the displacement chamber is shown on fig 2. When working as a pump, the intake stroke starts at the inner dead centre $V_{min}$ and terminates at the external dead centre $V_{max}$. There the connection with the intake tube is interrupted. The displacement chamber then becomes smaller again and the pressure increases depending on compressibility, as shown in fig 2, neglecting any leakage.

When the liquid in the displacement chamber has reached the delivery pressure, the connection with the delivery port is established by the slider and the liquid is expelled by movement of the piston towards the internal dead centre. The displacement chamber is disconnected from the delivery port and the pressure drops with the beginning volume increase depending on compressibility. This pressure drop is steeper, since here only the smaller volume $V_{min}$ is expanded.

Finally, the displacement chamber is again connected with the intake port. /FN[1b]/

FN[1b] — It is interesting to note, that the working fluid changes its thermodynamic state only during the compression and expansion parts, whilst the practically far more important intake and delivery parts correspond to a displacement of fluid in constant state.

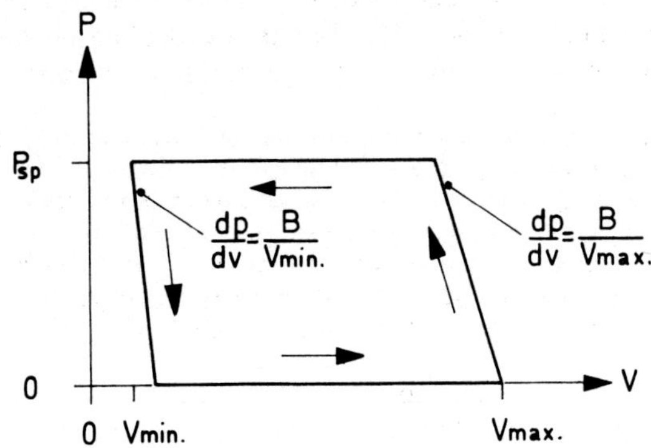



Fig. 31.2:  **Working  diagram  of  a  displacement  chamber**

In  order  to  obtain  the  described  ideal  working  diagram, it  is
necessary  to  open  the  connections  with  the  intake  and  delivery
ports  at  exact  moments. With  non-return  valves  in  both  ports, correct
timing  is  obtained,  but  only  at  low  rotation  frequency  since  otherwise
the  mass  of  the  valve  body  introduces  delays.With delayed  switching,
the  pressure  continues  to  increase  or  decrease  along  the  steep  parts
of  the  diagram,  leading  to  pressure  peaks  or  to  cavitation  in  the
the  displacement  chamber. If  the  connection  is  established  too  early,
one  obtains  sudden  pressure  equalisation  with  losses  and  shock
waves  (see  Section  73). If  the  hydrostatic  machine  works  as  a  motor,
the  working  diagram  is  run  through  in  reverse.

The  switching  moment  with  slide  valves  depends  on  their  movement  and  is  called  *positive  distribution*. The  correct  moment  or
timing  of  switching,  important  for  the  working  diagram,  depends  on
the  pressure  and  on  the  piston  stroke  in  machines  with  adjustable
displacement. Errors  in  timing  are  compensated  sometimes  by  suitable  leakage  flows.

The  effective  volume  flow  is  reduced  a  little  by  the  compressibility  of  the  oil,  by  about  1%  at  100  bar,  compared  to  the  geometric
displacement  volume.  /FN²/  For  a  more  detailed  treatment  of  compressibility  effects  refer  to  B 14,  Section  12.4.

Hydrostatic machine s with  positive  distribution  have  the advantage
that  they  can  also  work  as  motors. It  is  only  necessary  that  the

FN² — Also called  swept  volume.

pressure in the delivery port becomes larger than the corresponding force on the piston, whence the pistons are pushed outwards and the machine starts moving in the reverse direction as a motor.

It is possible to visualise the operation of a hydrostatic machine like a game of rope pulling (tug of war) between two teams. If the force on the pistons or the torque on the shaft is less than what corresponds to the delivery pressure, the machine reverses, taking oil from the delivery port and driving the shaft. Between both directions of rotation there is, as in rope pulling, a small range with no movement due to friction, but this corresponds only to a pressure difference of about 2 bar (30 psi).

The last remark brings up the question of losses in hydrostatic machines which take the following forms:

1. The leakage losses, due to undesired flow through sealing gaps, as for instance between pistons and cylinders.
2. The torque losses due to the friction on various sliding surfaces and also due to the pressure drop within the ports.

We have explained here the working diagram of a piston in a cylinder, but it is valid also for other displacement chambers. The following differences should be remembered:

1. The piston movement and therefore the displacement volume in piston machines is a sine function of the angle of rotation

   of the shaft. With other machines, for instance with gear pumps, this function may be quite different.
2. The ratio between the largest and the smallest volume of the displacement chamber is about 5 with piston machines (with adjustable displacement axial piston machines this is true only for maximum displacement), and can be quite different with other hydrostatic machines. It can, in particular, become very large with vane machines making them suitable as vacuum pumps.

### 31–3  Fundamental Equations without Losses

We shall now set up the quantitative relations between the operating variables, ie between flow, rotation frequency, displacement setting, operating pressure and torque. We shall neglect here compressibility effects and also the inevitable losses; that will be taken up in Section 31–4.

The displacement volume or *displacement* is the most fundamental variable of a hydrostatic machine. It is the volume delivered on each revolution or each radiant of rotation of the shaft. It can be determined by measuring the volume delivered with a given number of revolutions with vanishing delivery pressure. It can also be found approximately from the geometric data, ie with piston machines, from the number of cylinders, the piston area and the stroke.

The displacement volume or displacement of a hydrostatic machine can be given as the volume delivered by each revolution of the shaft, or as the volume delivered by one radiant of rotation of the shaft, equal to $1/2\pi$ revolutions or to $57°$. We shall use here only the displacement per radiant which will be denoted by $q$ The use of the displacement per radiant has the advantage that the factors $2\pi$ (see for instance Bibliography 14, which uses the displacement per revolution) disappear, leading to a welcome simplification of equations and functional diagrams.

Many hydrostatic machines have an adjustable displacement. We shall represent this with the displacement setting $a$ as follows:

$$q = a \, q_0 \qquad\qquad a = 0 \ldots 1$$

Here $q_0$ is the maximum displacement obtained with a displacement setting of 1. The flow is obtained by multiplication with the rotation frequency $\omega$, expressed in radiants per second.

$$Q = \omega q = a \omega q_0 \tag{31-1}$$

In order to determine the ideal torque without losses we consider that the work expended in a rotation through a radiant is equal to the torque. The hydrostatic energy is then equal to the corresponding volume multiplied with the pressure. Equating we obtain

$$M = p \, q = a \, p \, q_0 \tag{31-2}$$

We now determine the hydrostatic and the mechanical power by multiplying the flow with the pressure and the torque with the rotation frequency, respectively

$$P = \omega p \, q = a \, \omega \, p \, q_0 \tag{31-3}$$

We note immediately that the maximum power is obtained with $a = 1$ and with maximum values of rotation frequency and pressure

$$Pap = \omega_{max} \, p_{max} \, q_0 \qquad\qquad (31\text{--}4)$$

where Pap = maximum power

It should be noted that our equations 1 to 3 are applicable to both directions of rotation and of flow, producing then negative values of $\omega$, M and Q. If the pressure p is taken as difference of the absolute pressures at intake and delivery, it may also become negative. In order to avoid cavitation, one must prevent the absolute pressure in any port becoming too small by means of suitable feeding valves, as explained in detail in B 14.

Many hydrostatic machines with adjustable displacement can deliver also in the reverse direction without a change of direction of the rotation. This can be described by admitting a variation of the displacement setting between minus one over zero to plus one, $a = \backslash\text{-}1 \ldots . 0 \ldots . . 1.$

Hydrostatic machines with positive distribution can also deliver in reverse by reversing the sense of rotation. On the other hand, hydrostatic pumps with distribution by non-return valves deliver still in the same direction with reversed rotation even if they are equipped with adjustable displacement. Therefore, our fundamental equations $1 - 3$ must be applied to them with some limitation, but we shall not treat this point any further.

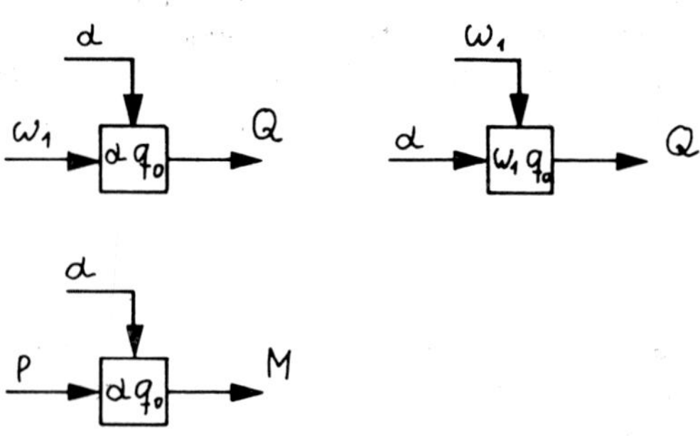

**Fig. 31.3: Functional diagram for hydrostatic machines without losses**

The functional diagram in fig 3 represents the equations 1 — 3. We find above the relations between flow, displacement setting and rotation frequency according to equation 1 and below the relation between pressure, displacement setting and torque according to equation 2. The rotation frequency is the input variable above at left with the displacement setting as perturbation on the side of the block, while above at right the displacement setting is drawn as input variable with the rotation frequency as perturbation. We shall use the latter variant in Section 62.

We would like to note the following in fig 3:
1. The operation of the machine is described by two blocks corresponding to equations 2 and 3.
2. Both blocks can be inverted without restriction meaning that we could also consider flow and torque as given and determine pressure and rotation frequencies from equations 2 and 3.

Equations 2 to 4 are the fundamental equations of hydrostatic machines without losses, simplified by expressing the rotation frequency in radiants per second and the displacement in volume per radiant, thereby eliminating the formerly frequent factor $2\pi$.

A similar simplification is obtained in electronics with alternating current, where one uses the angular frequency instead of the ordinary frequency, thereby avoiding the factor $2\pi$ in many equations.

Many catalogues contain the displacement volume per revolution which we shall denote as usual by the letter q. In order to avoid confusion we use for the displacement per radiant, called in the following simple displacement, the same letter but with a small cross line.

$$\text{Thus we have } \bar{q} = \frac{q}{2\pi} \quad \text{and } \bar{q}_0 = \frac{q_0}{2\pi}$$

According to the author's experience it is most convenient to convert all data first to the units used here. We would only like to remind the reader here that the rotation frequency in radiants per second is approximately 10% of the number of revolutions per minute, as treated in Chapter 1.

**Numerical Example of an Axial Piston Machine**

The usual size of an axial piston machine as a displacement of

$$\bar{q}_0 = 19 \text{ cm}^3/\text{rd} \quad \text{or } q_0 = 119 \text{ cm}^3/\text{rev}$$

and a standard frequency of 150 rd/sec. We obtain from equation 1

$$Q_0 = \omega q_0 = 150 \ \frac{rd}{sec} \ \ 19 \ \frac{cm^3}{rd} = 2850 \ \frac{cm^3}{rd}$$

and the ideal (without losses) torque with 100 bar and $\alpha = 1$.

$$M_i = P q_0 = 19 \ \frac{cm^3}{rd} \ \ 100 \ \frac{daN}{cm^2} = 1900 \ daNcm = 19 \ daNm$$

From the above we obtain the following rule: *The torque in daNm with 100 bar pressure is equal to the displacement in cm³/rd.* This can be written as follows:

$$\frac{M}{1 \, daNm} = \frac{P}{100 \ bar} \ \frac{q_0}{1 \ cm^3/rd} \tag{31–2b}$$

The power becomes with equations 3 and $\alpha = 1$.

$$p = 150 \ \frac{rd}{sec} \ \ 100 \ \frac{daN}{cm^2} 19 \ \frac{cm^3}{rd} = 27500 \ \frac{daNcm}{sec} = 27.5 \ kW$$

We finally obtain the following formula for practical determination of the power

$$p = \frac{\omega}{100 \ rd/sec} \ \frac{p}{100 \ bar} \ \frac{q_0}{1 \ cm^3/rd} \tag{31–3b}$$

*The power in kW is equal to the rotation frequency expressed as multiple of 100 rad/sec multiplied with the pressure as multiple of 100 bar and the displacement expressed as multiple of 100 rd/sec.*

With a maximum pressure of 300 bar (4300 psi) and a maximum frequency of 240 rad/sec (2500 rpm) we obtain for the maximum power of our machine by using equation 3b.

$$P_{ap} = 1 \ kW \ \frac{240}{100} \ \frac{300}{100} \ \frac{19}{1} = 135 \ kW$$

In industry one uses normally machines with a displacement volume of $3 - 60$ cm³/rd (or $20 - 400$ cm³/rd), corresponding to flows of 30 to 600 lit/min (6.5 to 130 gpm) at 150 rd/sec.

### 31–4   Fundamental Equations with Losses

So far we have neglected the inevitable losses and set up our equations 1 to 3 for ideal loss-free hydrostatic machines. Sometimes we affix the letter i, to the symbols of flow and torque to indicate that they are ideal values. Now we shall examine how the fundamental equations have to be modified to account for the losses in general, whilst we shall treat in the next subsection the losses more precisely and give their relation with mathematical models. The losses manifest themselves as a leakage and as a loss torque due to the various frictions in the hydrostatic machines.

The leakage is composed of flows through the many sealing gaps in hydrostatic machines. It should be proportional to pressure and inversely proportional to the viscosity as we have seen in Chapter 2. However, some deviations exist, and leakage is usually a function increasing more than in proportion to the pressure. On the other hand, it is proportional to the maximum displacement for a series of similar hydrostatic machines. Leakage can therefore be represented either by a *leakage admittance* $Y_S$, or by a *slip frequency* $\omega_S$, as defined by the following equation:

$$Q_S = Y_S\, P = \quad . \quad \omega_S\, q_0 \tag{31–5}$$

The leakage admittance is thus proportional to maximum displacement, and the slip frequency approximately proportional to pressure, but both increase with shaft rotation frequency.

The torque losses are mostly produced by the friction at the many sliding surfaces, but also from the natural pressure drop of the fluid within the machine. The torque loss is most conveniently represented by referring to the maximum displacement volume with a fictitious *loss pressure* as follows:

$$M_L = P_L\, (a_1\, \omega_1\, p)\, q_{0} \tag{31–6}$$

As we shall further discuss in Section 31–5, the loss pressure is practically independent of the maximum displacement and, therefore, the loss torque proportional to the latter with given values of rotation frequency, pressure and displacement setting. In many hydrostatic machines the loss pressure has a range of variation between 2 and 8 bar (30 to 120 psi).

As an indication, fig 4 gives the slip frequency of an axial piston machine as function of the pressure with the rotation frequency as

parameter. /FN³/ We see that it increases a little more than proportionally with the pressure, and also somewhat with the rotation
frequency.

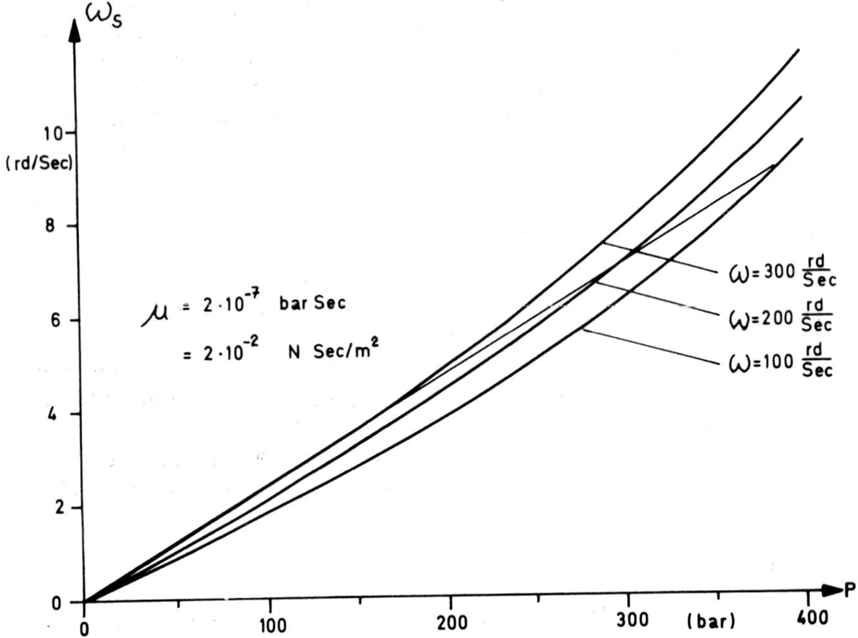


Fig. 31.4: Slip frequency of an axial piston machine as
function of operating pressure with rotation frequency
as parameter and with constant viscosity at the
entrance

The loss pressure of the same machine is represented on fig 5 as
a function of the rotation frequency and with the operating pressure
as parameter, with a displacement setting of 1. Fig 6, on the other
hand shows /FN⁴/ the increase of loss pressure as function of the
displacement setting at a given rotation frequency and operating
pressure. These curves are drawn from measurements on an axial
piston machine with relatively low losses, but they are representative of what can be obtained by careful design. In general, both
loss components are increasing functions of the operating variables.

FN³ — The broken line in fig 4 refers to the leakage coefficient determined later in
Section 31—4.

FN⁴ — The vertical arrows in figs 5 and 6 show the determination of the loss coefficients in Section 31—5.

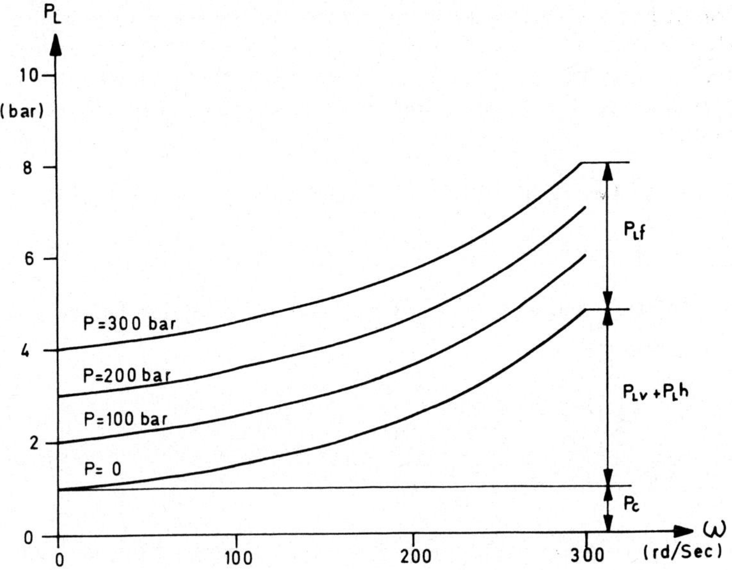


Fig 31.5: Loss pressure with function of rotation frequency for different operating pressures

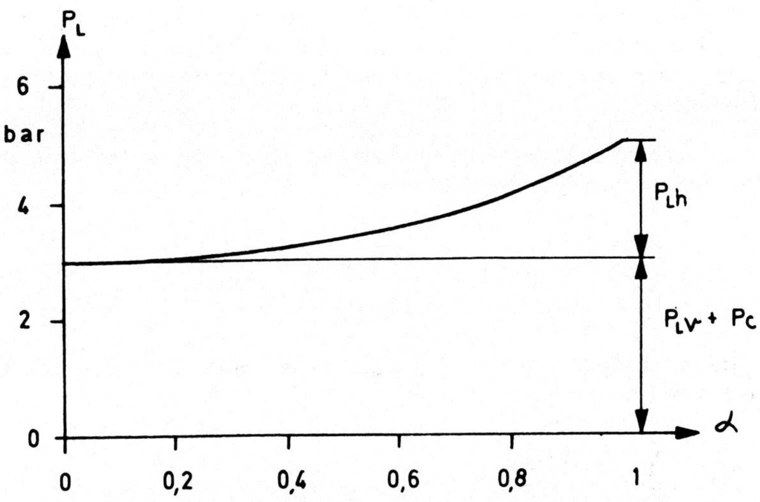


Fig 31.6: Loss pressure as function of displacement setting for constant rotation frequency and operating pressure

We would now like to set up equations for the effective torque and flow of hydrostatic machines with losses. In the case of pumping, we have to deduct the leakage flow and to add the loss torque from its ideal values, whilst in the case of motoring machines we have to add the leakage flow and deduct the loss torque. This gives for pumping:

$$Q_e = a \omega q_0 - Y_s \, p = a \omega q_0 - \omega_s q_0 \tag{31-7}$$

$$M_e = a \, p q_0 + p_L \, q_0 \tag{31-8}$$

The effective mechanical power $P_{me}$ and the effective hydrostatic power $P_{he}$ becomes

$$P_{me} = \omega M_e = a \omega p q_0 + \omega p_L q_0$$

$$P_{he} = p Q_e = a \omega p q_0 - \omega_s p q_0 \tag{31-9}$$

Since in motoring hydrostatic machines, leakage flow is added and loss torque deducted, the signs $+$ and $-$ have to be exchanged in their case.

The concepts of loss pressure and slip frequency are in the author's opinion more suitable for applications and for calculation of effective performance than the more usual efficiency. However, for completeness we shall also give here the volumetric and mechanical efficiencies.

In pumping machines, the volumetric efficiency is the ratio of effective to ideal flow and becomes:

$$\eta_v = \frac{Q_e}{Q_i} = 1 - \frac{\omega_s}{a \omega} \tag{31-10}$$

The mechanical efficiency is equal to the ratio of ideal to effective torque (in a pumping machine):

$$\eta_m = \frac{M_i}{M_e} = \frac{a \, p q_0}{M_e} = \frac{1}{1 + p_L / a \, p} \tag{31-11}$$

The total efficiency is defined as the ratio of hydrostatic to mechanical power:

$$\eta_{tot} = \frac{P_{he}}{P_{me}} = \frac{a\,\omega\,p\,q_0 - \omega_s\,p\,q_0}{a\,\omega\,p\,q_0 + \omega\,p\,q_0} \qquad (31\text{--}12)$$

$$\eta_{tot} = \frac{1 - \omega_s/a\omega}{1 + p_L/a\,p} = \eta_v\,\eta_m$$

For machines with fixed displacement we have simply to set $a = 1$ in the above formulae.

The distinction between mechanical and volumetric efficiencies can be made properly only with incompressible fluids. In practice we first determine the ideal flow by equation 7 from the effective flow with vanishing operating pressure. In actual measurements, it may perhaps be difficult to produce exactly vanishing operating pressures; therefore we extrapolate to a pressure of zero from the graph of effective flow versus operating pressure. Dividing by the rotation frequency gives the displacement. /FN[5]/ Having determined the displacement, we can find the ideal torque with each pressure by equation 1 and then the loss torque by measuring the effective torque, leading to the loss pressure and to the mechanical efficiency. Measuring the effective flow gives the slip flow by deducting the ideal flow determined beforehand, and thus the slip frequency.

For accurate measurements it is important to indicate at what pressure the effective flow is measured, due to the finite compressibility of the fluid. In principle, it is best to define the effective flow as the volume flow at one-half operating pressure, because it has then the correct energy content due to linear volume reduction with applied pressure. Otherwise the compressibility can be treated as an additional leakage, as described in B.14.

This slight indetermination of the effective flow comes from the finite compressibility and is outside of strict hydrostatics. Finally, it is a question of the precise definition of the terms.. On the other hand, when measuring the overall losses of hydrostatic transmissions we encounter no such difficulties, because the mechanical input and output power are both given as the product of rotation frequency and torque of the appropriate shaft, having no doubt or imprecise definition.

FN[5] — The author prefers this procedure to the CETOP recommendation for defining and measuring the displacement, although the difference is sensible only when very high accuracy is desired. Furthermore, newer work at the Technical University of Eindhoven indicates that a definition of the displacement volume similar as above, namely as the derivative of the flow with respect to rotation frequency under vanishing operating pressure would be preferable. See the interesting publication : G. Toet, The Determination of the theoretical swept Volume, Report from the Laboratory for Power Drives, Technical University, Eindhoven, Holland.

### 31—5    Mathematical Models for Losses

We have treated the losses in the preceding subsection only empirically, and would now like to find how they behave as function of the operating variables. We shall correlate them with our knowledge of operation, in particular with the characteristics of the gaps treated in Chapter 2. The so-called mathematical models have been set up for this purpose.

A *mathematical model* is a set of concepts or ideas that allow us to determine how the operating variables depend on each other. /FN[6]/ In this sense, our loss-free fundamental equations 1 to 3 are already mathematical models, perhaps of zero order. In a more restricted sense mathematical models explain the behaviour of the losses, normally by appropriate formulae, and in particular the behaviour of loss pressure and slip frequency. The formulae of mathematical models contain a combination of power products of the operating variables, together with dimensionless coefficients.

We shall treat here the basic concepts behind the mathematical models of Wilson (B16), Schloesser /FN[7]/ and Thoma (B14).

*Leakage*

The actual leakage flow is composed by the contributions of many gaps each obeying, in principle, the gap formulae 22—4. If the gap height is proportional to the linear dimensions of the machine, the leakage becomes proportional to the displacement. Therefore, Wilson represents the leakage as follows:

$$Q_s = C_s \frac{p}{\mu} \, q_0 \qquad\qquad \omega_s = C_s \frac{p}{\mu} \tag{31—13}$$

The above leakage is governed by the viscosity and independent of inertia of the fluid. In his refinement Schloesser adds a leakage component controlled only by inertia forces, thus proportional to the root of the pressure

$$Q_{st} = C_{st} \left[ \frac{2\,p}{\rho} \right]^{1/2} q_0^{2/3} \tag{31—14}$$

FN[6] — The expression 'functional model' would be more appropriate, but we shall use here the accepted terminology.

FN[7] — Measurements on displacement pumps, Dissertation Technical University, Delft, 1959.

In Schloesser's model, the total leakage is obtained by addition of equations 13 and 14. The author considers this not convincing because a leakage proportional to the root of the pressure is contrary to the behaviour of most hydrostatic machines, where it increases *more* than proportionally to pressure. Trying to fit this formula to actual leakage curves in order to determine the coefficients, the leakage component $Q_{ST}$ would become negative. This is not possible because it would imply that $Q_{ST}$ should flow upwards, ie from low to high pressure parts. Furthermore, the inertia pressure drop should be in series with the viscous pressure drop as we have seen in Section 24—2. Finally, with any reasonable gap height, the mean velocity is very small and the consequent dynamic pressure negligible as we have seen in Section 22. The mathematical model of Thoma does not contain the leakage component controlled by inertia forces.

*Torque losses*

The torque losses come mainly from friction effects as follows:

*Dry or contact friction*   Here the friction force is considered independent of the sliding velocity, for instance, due to metallic contact, but proportional to the normal force or operating pressure and to the displacement of the hydrostatic machine. This gives with a dry friction coefficient

$$M_f = C_f \, p \, q_0 \tag{31—15}$$

The friction forces of underbalanced hydrostatic bearings in Section 25—8 behave like this loss component.

*Viscous friction*   This component comes from overbalanced bearings and gaps. It is proportional to viscosity and rotation frequency and, furthermore, to the areas and also the radius, allowing to write with the viscous coefficient

$$M_v = C_v \, \mu \, \omega \, q_0 \tag{31—16}$$

*Constant friction*   Wilson adds a term of constant loss torque representing seal and similar friction

$$M_c = p_c \, q_0 \tag{31—17}$$

The torque loss of Wilson's model is composed by the above loss terms. On his side, Schloesser adds a term due to inertia forces in the fluid, including churning of the oil by moving parts in the casing which seems well justified.

$$M_h = C_h \, \omega^2 \rho \, q_0^{5/3} \tag{31—18}$$

*Hydrodynamical losses.* The mathematical model of Thoma is set up especially for variable displacement hydrostatic machines. It neglects the influence of displacement variation on dry and viscous friction, due partly to variable piston stroke. The turbulent pressure drop in the ports and ducts of the hydrostatic machine is proportional to the square of the flow velocity or to $\alpha^2$, whilst their effect is proportional $\alpha^3$, giving a hydrodynamical loss torque as follows:

$$M_h = C_h\, \alpha^3 \omega^2 \rho\, q_o^{5/3} \tag{31-19}$$

This model has the advantage of accounting, if only approximately, for the variation of loss torque or loss pressure with the displacement.

Composing the different loss contributions produces the global losses or more precisely the loss pressure and slip frequency in the different models. Table 1 contains the effective flow and torque according to the different models, obtained by inserting equations 13

### Table 31-1

#### Wilson

$$Q_e = \alpha \omega q_o - C_s \frac{p}{\mu} q_o$$

$$M_e = \alpha p q_o + C_f p q_o + C_\nu \mu \omega q_{o1} + p_c\, q_o$$

#### Schlösser

$$Q_e = \alpha \omega q_o - C_s \frac{p}{\mu} q_o' - C_{st} \left[ \frac{2p}{\rho} \right]^{1/2} q_o^{2/3}$$

$$M_e = \alpha p q_o + C_f p q_o + C_\nu \mu \omega q_o + C_h \omega^2 \rho\, q_o^{5/3}$$

#### Thoma

$$Q_e = \alpha \omega q_o - C_s \frac{p}{\mu} q_o$$

$$M_e = \alpha p q_o + C_f p q_o + C_\nu \mu \omega q_o + C_h \alpha^3 \omega^2 \rho\, q_o^{5/3} + p_c\, q_o$$

to 19 into equations 7 and 8. The following Table 2 gives the slip frequency and the loss pressure in the model of Thoma. /FN[8]/ It contains also representative values of the loss coefficients determined from the figs 4, 5 and 6.

## Table 31—2

$$\omega_s = C_s \frac{p}{\mu}$$

$$P_L = P_L f + P_L \nu + P_L h + P_c$$

$$P_L = C_f p + C_\nu \mu \omega + C_h \alpha^3 \omega^2 \rho \, q_o^{2/3} + P_c$$

$$C_s = 4.8 \cdot 10^{-8} ; \quad C_f = 10^{-2} ; \quad C_\nu = 3.3 \cdot 10^4 ;$$

$$C_h = 4 \qquad\qquad P_c = 1 \text{ bar}$$

As an example of application, we shall determine the loss coefficients of the model of Thoma from the graphs on figs 4, 5 and 6.

Referring first to leakage, the model gives a slip frequency proportional to pressure independent of rotation frequency. Therefore, the curves of fig 4 are first approximated by the thin straight line going through the origin. Taking now a pressure of 200 bar, the straight indicates a slip frequency of 4.8 rd/sec, leading with $\mu = 2.10^{-7}$ bar sec to

$$4.8 \ \frac{\text{rd}}{\text{sec}} = C_s \ \frac{200 \text{ bar}}{2.10^{-7} \text{ bar sec}} \qquad C_s = 4.8 \cdot 10^{-9}$$

With the torque losses, the separation of the contributions of the different loss mechanisms to the loss pressure is a little more difficult. As indicated by the vertical arrows of fig 5, the loss pressure without operating pressure and with very slow rotation is due to the constant friction pressure. The increase with rotation frequency gives the sum of viscous friction $P_{L\nu}$ and

of hydromechanical $P_{Lh}$ loss pressure, and finally the increase with operating pressure the contact friction $P_{Lf}$.

FN[8] — The original model of B14 did not contain the constant loss pressure, but later measures have caused the author to include it.

We see easily from fig 5 that $p'_c = 1$ bar, and that the contact friction pressure is 3 bar for 300 bar operating pressure. Therefore,

$$3 \text{ bar} = C_f \, 300 \text{ bar} \qquad C_f = 10^{-2} \qquad p_c = 1 \text{ bar}$$

In order to separate viscous friction from the hydrodynamic losses, we could extrapolate the almost linear beginning of the curves to higher rotation frequencies. With variable displacement machines it is preferable and more accurate to identify the hydrodynamic losses with the reduction of loss pressure when reducing the displacement setting $a$ at constant rotation frequency, as seen on fig 6, for $q = 16$ cm$^3$/rd with $\omega = 300$ rd/sec and

$p = 0$. The figure shows a $p_{Lh}$ of 2 bar.

We obtain with $a = 1$

$$2 \text{ bar} = C_h \, \omega^2 \rho \, q_0 \, ^{2/3} = C_h \cdot 0.52 \text{ bar} \qquad C_h = 4$$

since

$$\omega^2 \rho \, q_0 \, ^{2/3} = \left[ 300 \frac{\text{rd}}{\text{sec}} \right]^2 \, 900 \frac{\text{kg}}{\text{cm}^3} \left[ 16 \frac{\text{cm}^3}{\text{rd}} \right] = 0.52 \text{ bar}$$

We note furthermore again from fig 5, that the viscous friction contributes also 2 bar to the total loss pressure at 300 rd/sec. This gives

$$2 \text{ bar} = C_v \, \mu \, \omega = C_v \, 6 \, 10^{-5} \text{ bar} \qquad C_v = 3.3 \cdot 10^4$$

since $\mu \omega = 2.10$ bar sec. $300 \frac{\text{rd}}{\text{sec}} = 6.10^{-5}$ bar

The coefficients just determined are also entered on Table 2.

Naturally, this fitting process is by no means accurate, since with different values of viscosity and another range of operating variables we could obtain slightly different coefficients. The value of mathematical models is that they give some idea how the losses should behave in principle, especially for computer simulation. Their use gives definitely better accuracy than to neglect the losses altogether, as done only too often in industry.

More complicated models have been set up, for instance by Professor T. Iishihara, Tokyo, with more coefficients. However, in

ered. final analysis, the coefficients must be determined as done above, by
fitting the experimental loss curves, or the graphs of loss pressure
and loss frequency. They account only imperfectly for the observed
behaviour, and do not represent increase of leakage with rotation
frequency. Furthermore, the actual behaviour depends on many disturbances, including temperature. The losses in actual operation are,
therefore, not determined to a high accuracy. The author prefers to
use the graphs of slip frequency and loss pressure as given in figs
4, 5 and 6 for performance calculations. Nevertheless, mathematical
models are of great interest in principle, since they attempt to
establish some theoretical basis for the behaviour of hydrostatic
machines.

### 31 - 6   Flow Variations of Piston Machines

Most hydrostatic piston machines have single acting pistons
where the movement is a given function of shaft angle or time. The
delivery of one piston is then a rectified sine line, like a rectified
alternating current, since there is no delivery during the intake
stroke.

In order to reduce the variation of delivery flow it is usual to
provide several pistons and cylinders displaced each by the same
angle on the shaft. This is very effective especially with an odd
number of pistons since then the flows of the cylinders complement
each other relatively well. As an example the flow variation of a
three-cylinder machine is the same as of a six-cylinder one, and
similarly the variation of five cylinders the same as with ten cylinders.

**Table  31–3**

Flow Variations of Multicylinder Piston Machines

| $z$ = | 1 | 2 | 3 | 4 | 5 | 6 | 7 | 8 | 9 | 10 | 11 |
|---|---|---|---|---|---|---|---|---|---|---|---|
| $Q_{max} - Q_{min}$ = | 100 | | 13.4 | | 4.9 | | 2.5 | | 1.5 | | 1.0 % |
| $\dfrac{\phantom{Q_{max} - Q_{min}}}{Q_{max}}$ = | | 100 | | 29.3 | | 13.5 | | 7.6 | | 4.9 | % |

In order to study the compensation of flow variation of cylinders it is possible to represent them by phasors like in electronics. The phasors belonging to several cylinders form a regular polygon due to the constant angle between them. Their diameter represents the flow as function of the shaft angle, as further explained in Bibliography, 14

We shall not treat the flow variations any further here, but take from Bibliography, 14, Section 13, Table 31—3 with variations of the flow depending on the number of cylinders. The flow variation is referred not to the average, but to the maximum flow since it simplifies the calculation, but the difference is negligible with five or more cylinders.

## 32    Hydrostatic Piston Machines

### 32—1    Classification

After considering in the preceding section the operation of hydro-static machines in general, we shall study here the design and con-struction of the industrially used piston machines.

Similar to automobile engines, hydrostatic machines comprise a number of cylinders working with suitable phase differences in a cylinder block. This allows us to obtain a given displacement with small pistons and small piston velocity, and compensates at the same time the variations of the delivery as mentioned in Section 31—6. Contrary to automobile engines, the balancing of inertia forces of the moving parts is not important, since they are small compared to the forces of the working pressure.

We can classify piston machines as follows according to the form of the cylinder block:

1.  Machines with pistons in line.
2.  Axial piston machines.
3.  Radial piston machines.
4.  Other piston machines

From the point of view of the distribution of the liquid to the cylinders the machines can be classified as follows:

a.  Distribution by non-return valves in intake and delivery ports.
b.  Positive distribution by driven slides or ports.

Automobile engines have positive distribution in this sense, in four stroke engines by valves driven through a cam shaft and in two stroke engines by slits or ports in the cylinder walls.

Whilst hydrostatic machines suitable for motoring must have positive distribution, there are many hydrostatic pumps with non-return valves, mostly with conical valve bodies. Such valves, of the seating type as defined in Section 41, have the advantage that their leakage can be negligible.

Many piston machines have an adjustable displacement volume, in order to be able to change the delivery with constant rotation frequency, by one of the following methods:

1. Adjustment of displacement by variation of the piston stroke.
2. Adjustment of displacement by an additional outlet port from the cylinders.

The selection of the piston stroke can, in principle, be done just as with the slide valve of a steam locomotive, with a suitable rocking member, as shown in fig 11–1 of B.14. This method is frequently used with metering pumps in chemical engineering, but in hydrostatic engineering one uses today different methods as we shall see presently.

The other method of controlling the displacement uses an additional outlet port from the cylinder during a part of the delivery stroke. The fluid is expelled practically without pressure through it and losing only very little power, but a non-return valve in the delivery port is needed to prevent the oil in the delivery line to escape through the outlet port. On closing the additional outlet in the middle of the stroke the transition from low to high pressure in the cylinders takes place under high instantaneous piston velocity, leading to shocks and vibrations (see Section 71).

From the design point of view the additional outlet port can be made of a separately driven slider or of suitable slits in piston or cylinders. Fuel injection pumps of diesel engines are a good example of the second method, with control of the additional outlet port by rotation of the piston, with a recess in the form of a steep helix.

We shall include in the following a separate subsection for radial and axial piston machines, but for other design forms the following remarks may be sufficient.

*Machines with pistons in line.* Hydrostatic machines with pistons in line have a general layout similar to automobile engines, with cylinders on a parallel axis side by side in a block. The pistons are driven by connecting rods from a shaft, equipped usually with eccentrics and not with cranks, carrying roller bearings. Such pumps are built with distribution by non-return valve for very high pressure (400 bar or 6000 psi and more).

*Other piston machines.* The title *'Other piston machines'* comprises certain designs half-way between radial and axial piston machines. They have inclined cylinder axis, similarly as the spokes of a half open umbrella, but the author fails to recognise any merits.

## 32–2   Radial Piston Machines

The cylinders of radial piston machines are arranged in a normally rotating cylinder block in the form of a star, similar to the spokes of a wheel, as represented in fig 1. The pistons transmit the force over a ball joint for the compensating movement and over slippers on an external track ring. The eccentricity of the track ring with respect to the shaft can be adjusted and determines the piston stroke. The contact area between the slippers and the track ring is not plain, but rather a relatively flat part of a cylinder mantle; however, this does not essentially change the situation in the gap between the sealing lips and the ring. The sliding velocities are high due to the large radius, and it is advisable to use over-balanced hydrostatic bearings with lubricant supply through the pistons and ball joints by the oil ducts shown. Sometimes the oil duct of the piston axis is not extended until the cylinder space, but connected with a cross bore near to the head of the piston, providing the required impedance or throttling in the short gap between piston and cylinder.

**Fig 32.1:** Schematic     cross-section     through     radial piston machine

The fluid is transported by the short cylinder channels and ports in the rotating cylinder block into the not rotating pintle, carrying the intake and delivery ports. The cylinders register (ie contact) intermittently with the admission and the delivery ports, constituting thereby a positive distribution. This arrangement is generally called a pintle valve.

Radial piston machines have the following limitations:
1. Small cross sections of the ports in the pintle.
2. Difficult design of the pintle surfaces.
3. High sliding velocities of the slippers on the track rings.

In addition to the design on fig 1 there are many other forms of radial piston machines. One variant comprises a track ring suspended in suitable journal bearings and revolving, so that the slippers have to take up only compensating movements. Another design uses a not rotating cylinder block with positive distribution by mushroom-type seating valves and cams, similar to the ones used in automobile engines. The piston stroke is adjustable by two eccentrics, arranged one between the other, such that the relative angle can be changed.

There exists also the so-called low speed high torque hydrostatic machine which we shall treat in subsection 32—5. Many of them are made as radial piston machines, because with the low rotation frequency the limitations 1 and 3 above disappear.


### 32—3   Axial Piston Machines

The pistons of axial piston machines are arranged in a barrel-like, normally rotating cylinder block, parallel to its axis and evenly distributed over its circumference.  The pistons exert their forces on a plane inclined with respect to the axis of the cylinder block, producing a piston movement similar to what we can observe with automobile wheels on distorted axles. Whilst the inclined plane was driven originally by a universal joint, today the following types are generally used:

1. Axial piston machines with swash plate.
2. Axial piston machines with tilting head /FN[1]/

---

FN[1] — Called in USA usually 'bent axis machines'.

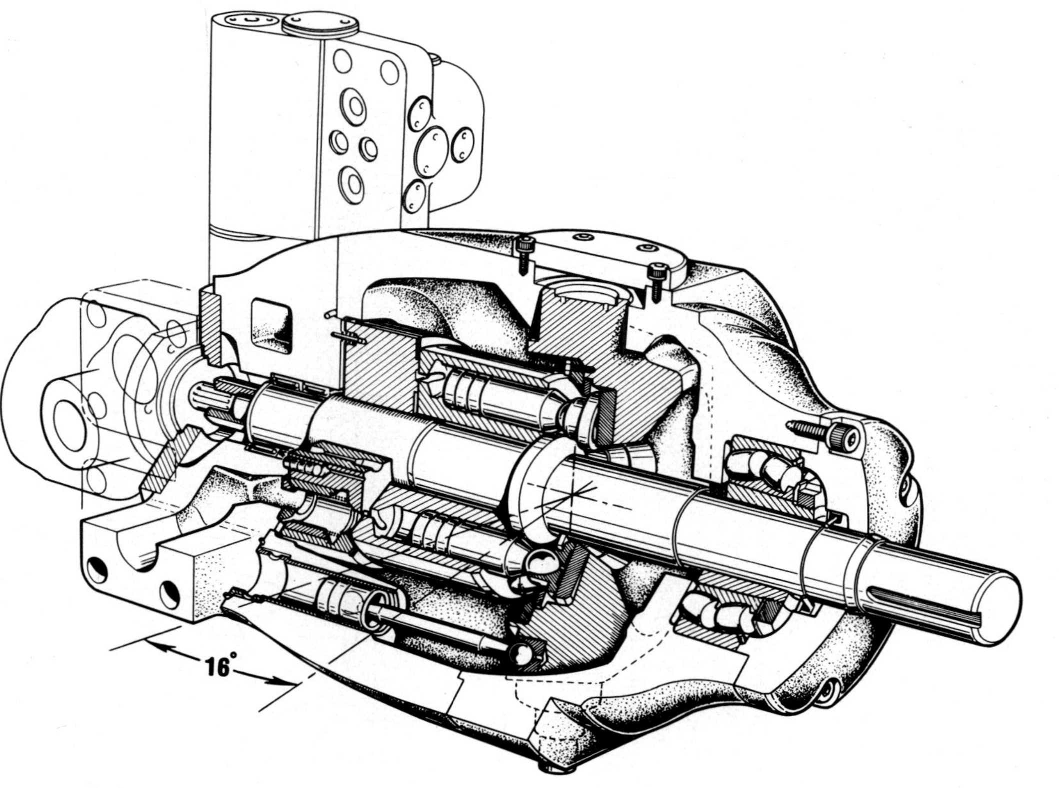


Fig 32.2: Swash plate machine with floating distributor and auxiliary pump at the left (VSG)

Fig 2 contains a perspective representation of a machine where the inclined plane is built by a non-rotating swash plate. This swash plate is therefore inclinable around an axis perpendicular to the axis of the shaft, and equipped with a track ring on which the pistons abut over slippers and ball joints. The admission and delivery of the fluid takes place over cylinder channels in the cylinder block, over the kidney-shaped timing ports in a fixed distributor or timing plate. The latter is connected over short, slightly adjustable pieces of tubing with the casing. Thus it maintains an intimate contact with the cylinder block in spite of small errors in its position. This design is called sometimes a floating distributor, as treated in Section 33–4.

We use fig 3 with a piston and a slipper in its centre position to explain the forces of swash plate machines. It is best to imagine that the swash plate exerts a force $F_{sl}$ on the slipper in a direction perpendicular to its surface, neglecting friction. This force is decomposed in the spherical head into two components, a piston force $F_p$ and a radial force $F_r$ as follows:

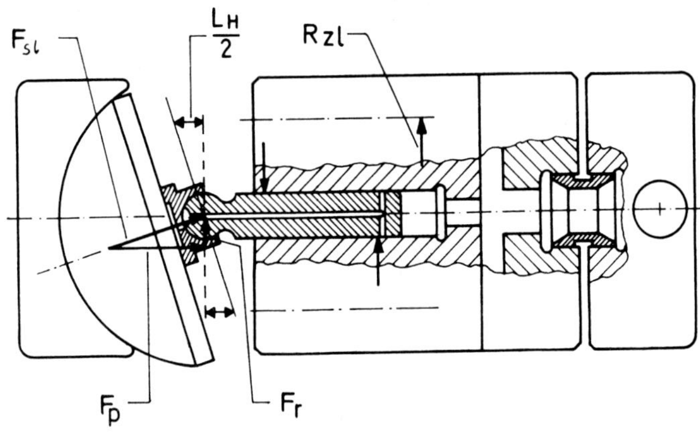


Fig 32.3: Schematic cross-section through swash plate
machine with indication of the forces

$$F_p = F_{sl}\ \cos\hat{\alpha} \qquad F_r = F_{sl}\ \sin\hat{\alpha} \qquad\qquad (32-1a)$$

With given piston force we obtain

$$F_{sl} = \frac{F_p}{\cos\hat{\alpha}} \qquad\qquad (32-1b)$$

The normal force or load of the slipper is a little larger than the piston force, about 5% with a tilting angle of 18°, whilst the radial force is about 32% of the piston force.

The radial force is loading the piston like a cantilever producing the reaction forces indicated by the vertical arrows of fig 3. The piston must penetrate sufficiently deep in the cylinder even at external dead centre, in order to limit the reaction forces. As a practical rule, the penetrated piston length at external dead centre should be at least 150% of the distance (at external dead centre) between the centre of the spherical head and the end of the cylinder block. This corresponds roughly to 2.4 piston diameters.

The reaction forces on the cylinder block attack in reality outside the plane of the paper, producing thereby the useful torque which is then transmitted from the cylinder block to the shaft.

On fig 3 we can easily see that the instantaneous position of the piston is given by

$$S_p = R_{zl} \tan \hat{a} \sin \phi \tag{32-2}$$

with $R_{zl}$ the pitch radius of the bores of the cylinder block, $\hat{a}$ tilting angle /FN1b/ and $\phi$ of the rotation angle of the shaft. The stroke then becomes

$$l_h = 2 R_{zl} \tan \hat{a} \tag{32-3}$$

A machine with z cylinders of diameter d has thus a displacement volume of

$$q = \frac{9}{2\pi} = \frac{z}{2\pi} \frac{d^2 \pi}{4} \quad 2 R_{zl} \tan \hat{a} \tag{32-4}$$

Equation 4 shows how the displacement is changed by adjusting the angle $\hat{a}$ It is valid also when tilting the swash plate in the inverse sense, leading to negative tilting angle and displacement volume. The latter means simply that the machine delivers a flow in reverse with the same direction of rotation.

---

FN1b — We use the symbol $\hat{a}$ for the angle of inclination, since $\hat{a}$ denotes the displacement setting defined in Section 31-1.

Numerical example : For a hydrostatic machine with the displacement mentioned in Section 31 one frequently uses pistons with dp = 24mm, A = 4.52 cm². $\hat{a}$ = 17° and with a pitch radius in the cylinder block of R zl = 48 mm. The stroke becomes from equation 3

$$L_h = 24.8 \text{ cm tan } \hat{a} = 2{,}92 \text{ cm}$$

The displacement volume becomes

$$q = \frac{9}{2\pi} \; 4.52 \text{ cm}^2 \; 2.92 \text{ cm} = 19 \text{ cm}^3/\text{rd}$$

The slippers run on a track with the radius of 4.8 cm and with a rotation frequency of 150 rad/sec. The sliding velocity becomes

$$v = 150 \frac{\text{rd}}{\text{sec}} \; 4.8 \text{ cm} = 720 \frac{\text{cm}}{\text{sec}}$$

We have calculated the oil consumption of 5.2 cm³/sec already in Section 25-1, and this contributes to the slip frequency as follows:

$$\omega_s = \frac{5.2 \text{ cm}^3/\text{sec}}{19 \text{ cm}^3/\text{rd}} = 0.27 \frac{\text{rd}}{\text{sec}}$$

The friction force of the slippers or the corresponding loss torque produces a loss pressure of

$$p_l = \frac{4.8 \text{ cm } 3.8 \text{ daN}}{19 \text{ cm}^3/\text{rd}} = 1 \frac{\text{daN}}{\text{cm}^2} = 1 \text{ bar}$$

The slipper thus produces

$$\frac{0.27 \text{ rd/sec}}{160 \text{ rd/sec}} = 0.18\% \text{ at the volumetric efficiency}$$

$$\frac{1 \text{ bar}}{100 \text{ bar}} = 1\% \text{ of the mechanical efficiency}$$

The losses of the other components have to be added.

The retraction of the pistons is a difficult design detail. It can be made by a retracting plate contacting the slippers like a fork above the lower enlargement, visible on fig 3. It is used only to retract the pistons in extraordinary cases, like insufficient priming pressure.

There exist many different designs of swash plate machines. Some of them do not use slippers, but the pistons are equipped with ball ends directly abutting against the track ring.This ring is then usually suspended over roller bearings in such a way that the ball ends have to take up only compensating movements. Nevertheless, they have a Hertzian contact on the track ring similar to ball bearings, producing the same relation between load or operating pressure and expected life (life increases with the third power of the reciprocal load). By suitable selection of the radius of the ball ends, the pitch radius and the tilting angle, the ball ends roll on the track ring without sliding (see B.8, page 34). Fig 54—7 below shows a schematic cross-section through a hydrostatic machine as servomotor with a rotating servovalve incorporating this piston design.

## Tilting Head Machines

Tilting head machines incorporate an inclined plate where connecting rods are fixed by ball joints without slippers. The connecting rods can exert forces in the plane of the inclined plate, contrary to the swash plate design, since their position is determined by that of the spherical sockets.

In order to accommodate the lateral movement it is necessary to anchor the connecting rods with a further spherical joint in the piston, but they have to take up only a very few degrees.

The inclined plate forms today part of the drive shaft in the form of rigidly fixed drive flange, tilted with respect to the axis of the cylinder block. Normally it is preferred to have a drive shaft fixed in space, and to use an inclinable cylinder block in a tilting head around a transverse axis. The tilting axis is situated in the plane of the ball joints of the drive flange. For the admission and the delivery into the tilting head one uses suitable sleeve bearings within the tilting trunnions.

Fig 4 shows a perspective drawing of a tilting head machine with visible bores for the fluid and conducting sleeves in the tilting trunnions. The drive shaft is suspended by ball bearings in the casing. This figure shows the so-called *wet mounted type*, also called *flange type,* which must be built inside an oil tank, since some leakage will come out from the rear side (at right). The body of the machine would have to be completed by a casing and a rear cover in order to allow open mounting.

Fig 5 shows another variable displacement tilting head machine, with a displacement of 16 ccm$^3$ per radiant, together with an oil tank

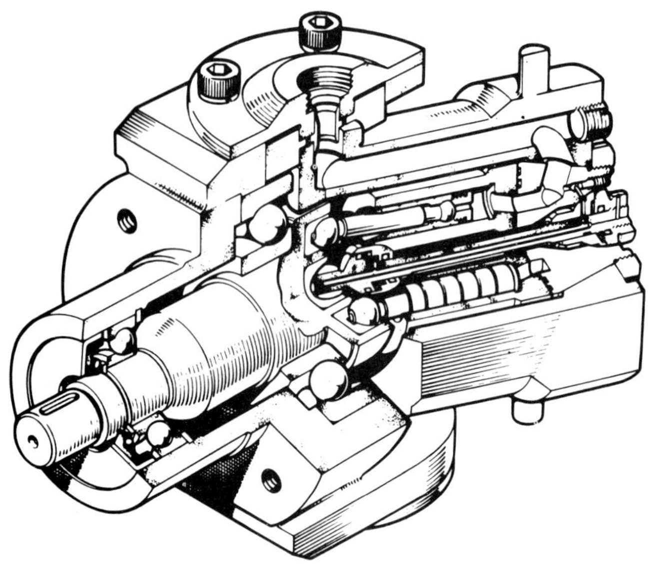

Fig. 32.4: Flange-mounted tilting head axial piston
machine (Cattermole)

and an electric drive motor. This machine is equipped with an auxiliary pump for priming oil, and with a number of valves as needed
for hydrostatic machines in the closed circuit (see Section 72). This
machine is built into a sealed casing and can be used anywhere. The
devices necessary for adjusting the displacement are found in the
towerlike structure.

We shall use fig 6 with a schematic cross section perpendicular
to the plane of the tilting axis in order to study the forces. We see a
cylinder and a piston in their central position, which is in reality
offset by the pitch radius put off the plane of the paper. The piston
presses over the small spherical joint and the connecting rod on the
large spherical joint in the drive flange. This force is decomposed as
follows in axial and radial force in the drive flange

$$F_a = F_p \cos \hat{a} \qquad F_r = F_p \sin \hat{a} \qquad (32-5)$$

The radial force produces with its offset out from the plane of the
paper the useful torque of the shaft.Contrary to swash plate machines,

Fig. 32.5: Tilting head machine in casing with electro-
hydraulic displacement setting in the towerlike
structure (Galdabini)

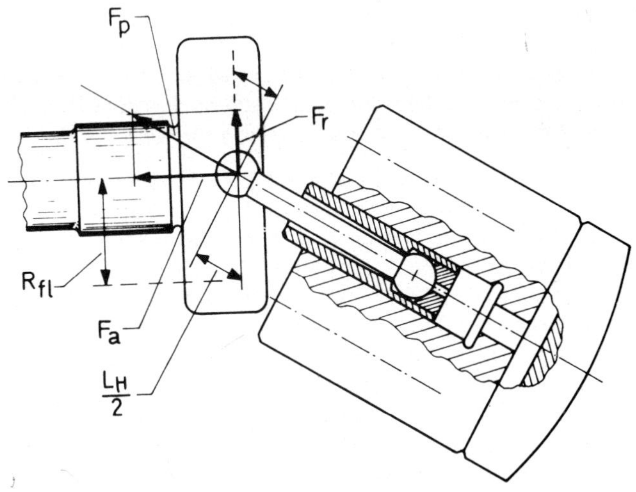


Fig. 32.6: Schematic cross-section of a tilting head
machine with indication of the forces by arrows

there are no large side forces on the cylinders, the remaining forces corresponding only to the inclination of the connecting rod with respect to the cylinder axis (see Section 33—6).

The stroke of the piston becomes, as we can see on the lower half of fig 6

$$L_h = 2 R_{fl} \sin \hat{a} \qquad\qquad (32\text{--}6)$$

and therefore the displacement volume

$$q = \frac{9}{2\pi} = \frac{z}{2\pi} \frac{d^2 \pi}{4} 2 R_{fl} \sin \hat{a} \qquad\qquad (32\text{--}7)$$

Here $R_{f1}$ denotes the pitch radius in the drive flange which is usually about 5% larger than the pitch radius in the cylinder block (See Section 33—6).

Since there is no side force on the pistons, it is possible to build tilting head machines with large inclination angles. This gives a better angle of attack of the piston forces and smaller loss torques, especially with high pressure and low rotation frequency. Whilst the tilting angle is limited in swash plate machines to about 17 to 18°, many tilting head machines have 25°, sometimes even 35°, if the synchronisation between drive flange and cylinder block is made by a universal joint. One limiting factor of the tilting angle is the rapidly increasing compensating movement on the connecting rods (proportional to $\cos \hat{a}$), because it obliges us to make the connecting rods rather thin, whence they become liable to buckling.

In fig 7 we see a tilting head axial piston machine produced with a range of displacements from 0.8 to 12.5 cm/rd, with a maximum tilting angle of 45°. This design uses spherical pistons, manufactured integrally (from one piece) with the connecting rods. In order to increase the sealing length, the pistons are equipped with piston rings, resulting in sealing lengths of 12—14% of the piston diameter. /FN²/ Otherwise tilting head machines with synchronisation through a universal joint have an average piston length in the cylinder of about 1.5 piston diameters, and with synchronisation by connecting rods of about 2.5 piston diameters. In the above design the piston rings are subject to the operating pressure, reducing the gap height between them and the cylinder bores.

$FN^2$ — Spherical pistons without piston rings have sealing lengths of about 2.5% of the piston diameter, as we have seen in Section 23—5.

Fig. 32.7: Tilting head machine with 45° tilting angle,
synchronisation by bevel gears and spherical pistons
with piston rings (Flygmotor)

Comparing the relative merits of swash plate and tilting head axial piston machines, we can conclude : Tilting head machines have the following advantages compared to swash plate machines:

1. The friction losses on the transmission of radial forces are smaller due to the increased tilting angle.

2. The distributing surfaces can be designed more favourably, ie with less leakage and less loss torque at a given gap height, since they do not have to be built around the drive shaft.

3. Higher rotation frequencies are possible due to the smaller mass of and due to the absence of lateral forces on the pistons.

4. It is possible to anchor the ball joints well enough allowing a self-priming operation up to the limit due to cavitation. Such self-priming operation is usually possible with specific speeds up to 1.2 (see Section 35).

5. Since they can be built with larger gap heights, they are less sensitive to contamination of the oil.

On the other hand, tilting head machines have the following disadvantages:

1. They are somewhat more bulky, since space for tilting the head must be provided.
2. For rapid tilting more torque is needed, since the moment of inertia around the tilting axis is larger.
3. It is not possible to have through-going shafts or two shaft connections.
4. The operating pressure is limited in many designs by the load capacity of the roller bearings, but this can be avoided by hydrostatic suspension of the drive shaft.


## 32-4   Slow Speed Machines

Many applications demand a drive at low frequency but with a high torque, as can be produced by a normal hydrostatic motor followed by a fixed gear reducer. This increases the torque, but imposes the mass, bulk and cost of the gear reducer.

Another solution is to use a hydrostatic machine giving a high torque with limited rotation frequency, usually up to 25 rad/sec (250 rpm). Since the relation between torque and displacement is always given by our equation 31-2, the problem is to design a hydrostatic machine as small and as cheap as possible and suitable for large displacement, whilst renouncing high rotation frequency. Such machines are built in practice in a range of 100 $cm^3$/rad to 1200 $cm^3$/rad (600–7500 $cm^3$/rev), but normally only with |fixed displacement for simplicity. The maximum frequency expressed in the specific rotation frequency of Section 35 is about 0.5 corresponding to 20 rad/sec with 400 $cm^3$ per rad. The pressure limit is usually somewhat lower as with normal hydrostatic machines, in the range of 200 bar (3000 psi).

The above specification and operating limits have the following influences on the design:

1. The cross sections of the ports and ducts can be relatively small, since the flow at a given displacement is also small.
2. Leakage must be more limited in order that the slip frequency should not exceed a certain fraction of the rotation frequency.
3. The sliding velocities are small.

Underbalanced hydrostatic bearings correspond well to these specifications since they produce little leakage with the metallic contact of the sealing lips. Slow speed machines are also sometimes equipped with piston rings in order to improve sealing.

Low speed hydrostatic motors are produced in many forms, as axial piston, radial piston or vane machines (see 34—2), but at the present (1970) radial piston machines seem to be in the centre of interest. Due to the low rotation frequency the great disadvantage of this design, small cross sections for the flow, loses its importance.

Fig 8 represents an example of a radial piston machine with five cylinders. The shaft is equipped with an eccentric on which the non-rotating pentagon-like distributor is fixed. The pistons are abutting on its five flat end faces over the visible sealing elements.

The distributor piece and the pistons are equipped with large bores for the fluid, transmitting at the same time an important fraction of the forces by oil pressure. The shaft with eccentric is equipped with admission and delivery ports, and the distribution in the cylinder takes place between the eccentric and the distributor piece. The fluid is conducted from the left of the casing into the shaft by two annular grooves, one each for admission and delivery and separated by sealing rings. The transformation of the eccentric piston forces into a torque takes place in two conical roller bearings.

We note the following details on fig 8:
1. There are sealing rings with limited adjusting movements, in order to compensate for small alignment errors between the piston faces and the end faces of the pentagon distributor piece.
2. The hollow pistons expand under the internal operating pressure, improving the sealing, which is further assisted by piston rings.
3. The pistons are retracted by springs, a practice which is less doubtful at the low rotation frequency.

The variation of the torque with constant pressure is equal to the variation of the flow with constant rotation frequency, and therefore given by the values found on Table 31—3. It is objectionable sometimes with low rotation frequencies and this is why some other manufacturers prefer larger numbers of cylinders.

If the motor is not supplied with constant pressure, but from a power pack with given tangent admittance, there is a certain influence together with the characteristic of the load. The resulting

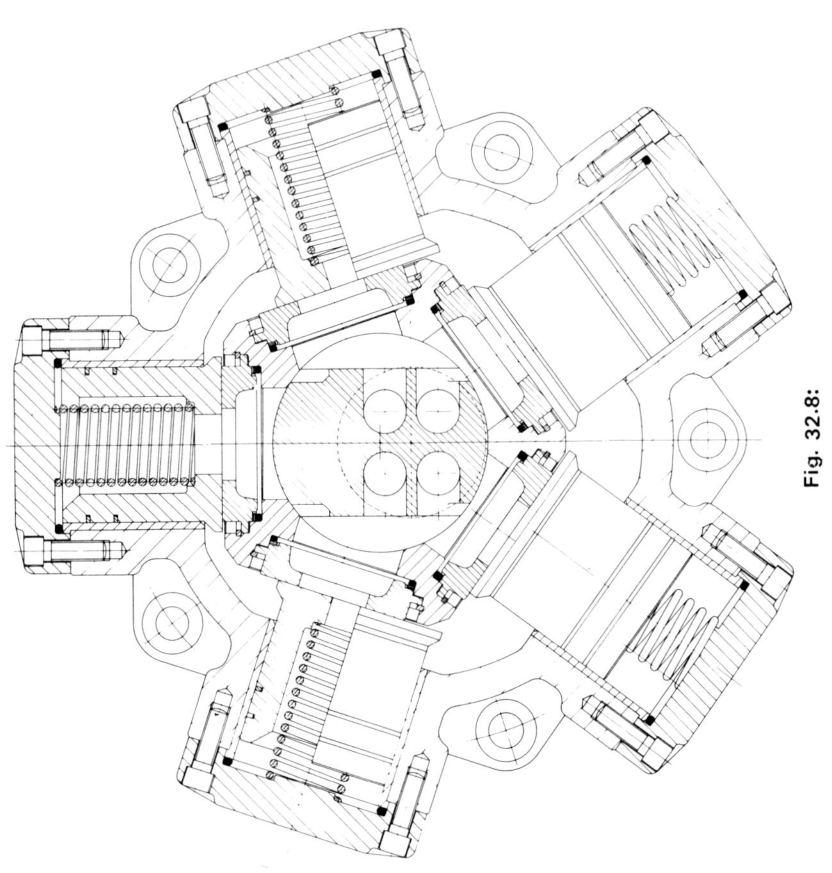

Fig. 32.8:

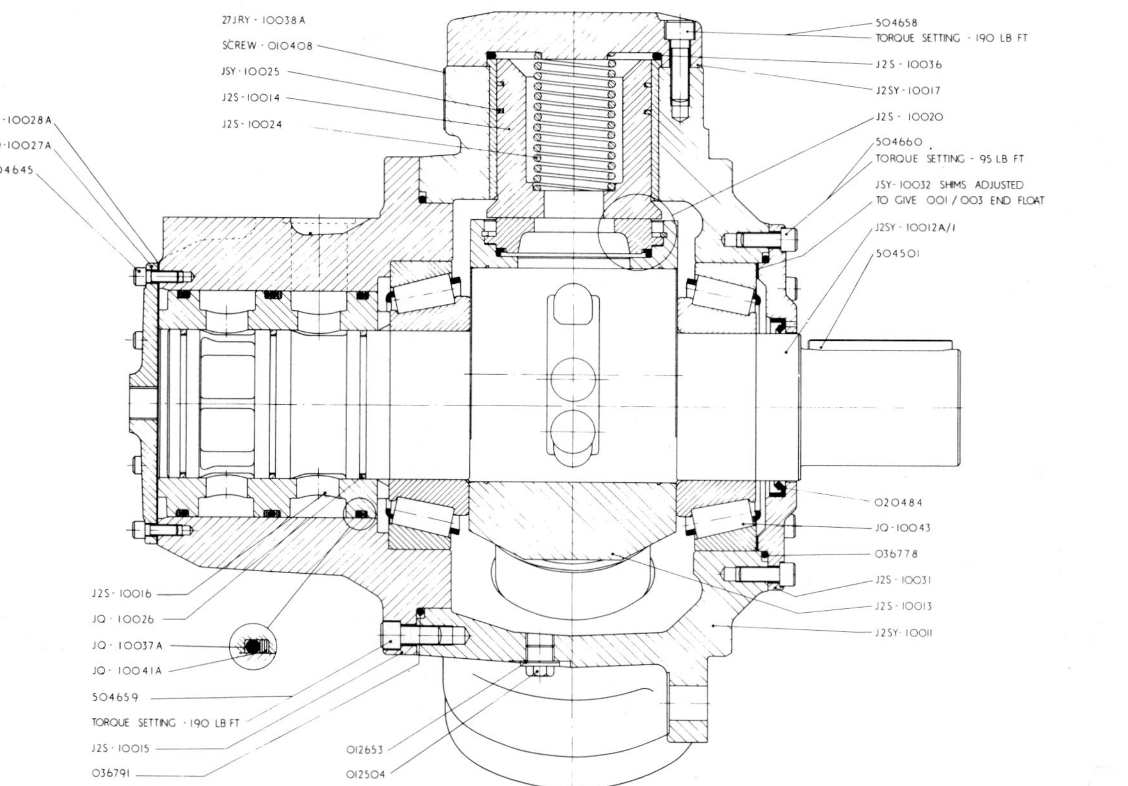


**Fig. 32.8: (Cont'd)** Longitudinal and cross-section through a low-speed radial piston machine (Ruston)

variations can be further studied with functional diagrams, but are beyond the scope of the present book.

Let it finally be mentioned, that even with slow speed motors, it is necessary to adapt the frequency to the requirements of the load. Frequently gear reducers are still needed, perhaps only with one stage, working at low speed. The reduced noise of such installations is a further advantage.

## 33    Component Parts of Piston Machines

After discussing, in Section 32, the design of hydrostatic piston machines in general, we shall now discuss the design of some important component parts. These are difficult problems, but their solution is facilitated by the fluid mechanics of the gap treated in Section 2, and by the application of geometrical principles.

The main difficulty in the design and determination of many component parts comes from the fact that hydrostatic machines must work satisfactorily despite variations of

1.  Rotation frequency,
2.  Operating pressure and
3.  Displacement setting

On the other hand, mass forces and all other forces, excepting pressure forces, are normally negligible.

### 33—1      Pistons with Slipper

We consider the piston with slipper, represented in fig 1, as a simple component part of an axial piston machine. The slipper is of the same form as discussed in Section 25, but it is equipped here with a concave, and the piston with a convex spherical surface. It is always desirable to build the slipper as low as possible in order that:

1.  The distance between the centre of the sphere and the end of the cylinder block becomes small, thus reducing the side forces on the pistons.
2.  The tilting moment on the slipper, given by the tangential force on the support multiplied by the distance to the centre of the sphere, should be as small as possible.

The slipper forms, together with its support, a hydrostatic bearing fed over a bore in the piston from the cylinder.

The throttling takes place in fig 1 in a short groove on the spherical head of the piston, and the oil is conducted through the eccentric bores of the slipper to the depression between the sealing lips. The

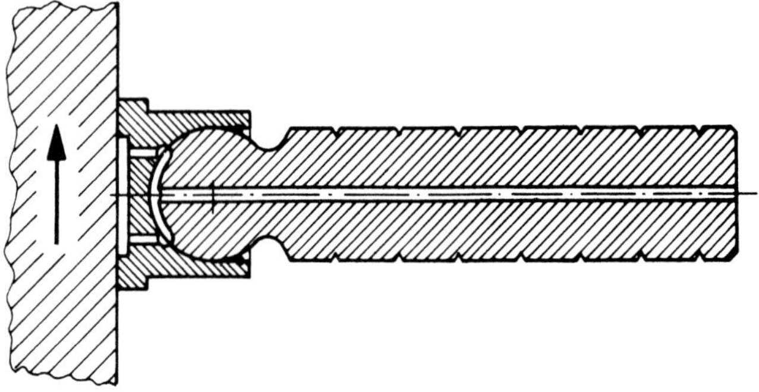

Fig. 33.1: Piston with slipper

throttling groove must have a certain length in order to avoid the movement of the slipper in the spherical joint, due to the tilting angle, which might produce a short circuit or change its impedance too much. The oil is collected by the annular groove in the spherical surface of the slipper compensating thereby for any rotation between slipper and piston. Unless prevented by suitable pins, such rotation always takes place, and is also desirable, since it equalizes loads and any wear.Pistons with slippers of this kind were used in fig 32–2.

Instead of the groove in the ball joint it is possible to use other impedances, including the throttling between piston and cylinder bores as shown in fig 32–1. Further designs have a throttling groove with the form of a spiral in an isle in the depression of the slipper, or simply a narrow bore, sensitive to clogging by impurities, between spherical seat and the depression. We shall describe the design and dimensioning of throttling devices in Section 32–2.

The security of the slipper depends on maintaining a sufficiently high gap with oil film at all operating conditions. We shall therefore study the gap height in analogy to fig 25–4.

The slipper is loaded by the piston force given by the piston area $A_p$ and the operating pressure, independently of the gap height. The operating pressure is also equal to the supplied pressure for the throttle. Therefore we obtain for the applied pressure $P_o$, that is the pressure in the depression of the slipper similar to equation 25–1 and to the numerical example of Section 25

$$P_o \; A_{tot} \; = \; P_{sp} \; A_p \qquad P_o = \lambda \, P_{sp} \qquad \lambda = \frac{A_p}{A_{tot}} \qquad 33\text{–}1$$

$$A_{tot} = A_{cv} + A_{lp}/2$$

Here the variable $\lambda$ is the load factor which we have already met in Section 25. It is given here by the ratio of the piston area to the active area of the slipper bearing.

In order to obtain the functional diagram fig 2, we only have to enter the load according to equation 1 into the former fig 25–4. The pressure drop in the throttle, equal to the difference between supply and applied pressure determines the flow according to the secant admittance $Y_{sc}$. It is divided in the next block by the applied pressure, and the following two blocks determine the third power of the gap height, and then the gap height itself.

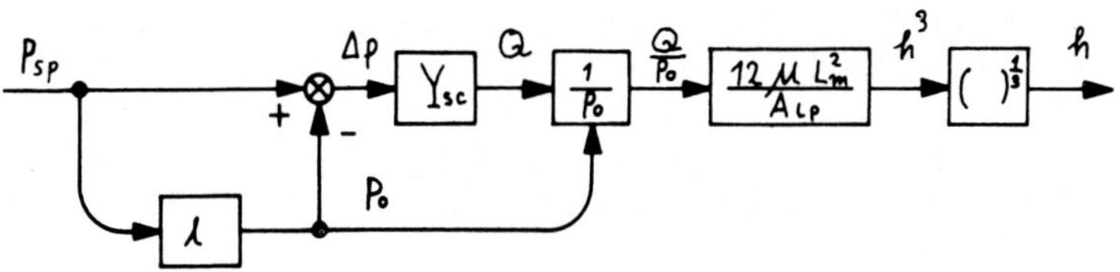

Fig. 33.2: Functional diagram for the gap height of a slipper loaded by a piston

The pressure drop is related to the load factor as follows:

$$\Delta p = P_{sp} - P_o = (1 - \lambda)\, P_o$$

The ratio between flow and applied pressure is, therefore, determined by the secant admittance of the throttle:

$$\frac{Q}{P_o} = Y_{sc}\, \frac{\Delta p}{P_o} = Y_{sc}\, \frac{\Delta p}{P_o} = Y_{sc}\, \frac{1 - \lambda}{\lambda} \tag{33–2}$$

The factor $\lambda$ in equation 2 depends (according to equation 1) only on the areas, independently of the operating conditions with given dimensions. Therefore, the ratio $Q/P_o$ and the gap height is determined principally by the secant impedance of the throttle.

The functional diagram in fig 3 shows how the secant impedance of the throttle determines the ratio $Q/P_o$ according to equation 2 and furthermore the gap height. It is in principle independant of the operating conditions, but in practice, subject to the following influences:

1. As we know from Chapter 1 the secant admittance is constant only with strictly linear throttle, but otherwise a function of the pressure drop, especially with turbulent conditions. This is represented in fig 3 by the perturbating connections arriving from left below at the block containing $Y_{sc}$.
2. The gap formula in the fourth block of fig 3 contains the mean viscosity $\mu$ rapidly variable with temperature. Depending on the heat conduction through the walls, the average temperature increases with operating pressure and sliding velocity, thereby reducing the mean viscosity.

The hydrostatic bearing built by the surface of the slipper is very useful in practice, since the gap height is approximately independent of the operating conditions in spite of these limitations. The necessary gap height is selected as in Section 25-6, but with very high pressures it may be desirable to reduce it in order to limit the oil consumption, always determined by the gap formula 22-4, or 25-5.

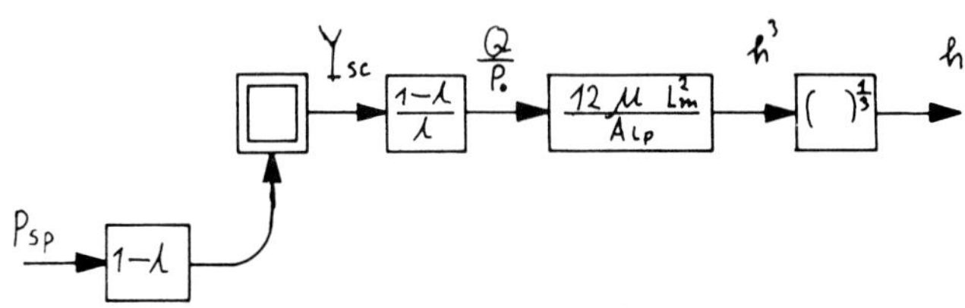

Fig. 33.3:   Modified functional diagram for the gap height
of a slipper with piston

On the other hand, it is desirable to increase the gap height with high sliding velocities and rotation frequencies, in order to improve the security against overheating and seizing, produced by a throttle with an admittance increasing with sliding velocity. According to the experience of the author, only the compression throttle (see Section 33-2) has these characteristics. It should also be mentioned that the load of the slipper depends on the tilting angle according to equation 32-1b, leading to a variation of the load factor, but there with the usual maximum tilting angle of 17°, the variation is only 5% and normally not disturbing.

### 33–2   Throttling or Metering Devices

Throttles or metering devices have an important function in hydrostatic bearings for stabilisation of the desired gap height. They are used very frequently and we shall treat their design here.

A throttle is a device having a required impedance or admittance. The difficulty is to make the throttling effect, or the impedance sufficiently large. A simple bore would have to be made very small for the usual impedances and therefore tends to be clogged by impurities in the oil.

Formerly there was the general rule that bores below 1mm in diameter should be avoided because of possible clogging. With the improvement in filtration, however, one today admits —especially with servo-valves — much more narrow bores, down to diameters of 0.15mm.

The danger of clogging is substantially reduced if the walls of the throttle have relative movement which can be realised as follows:

1.  The throttle is built as open channel or groove in a sliding surface, and covered by the other sliding surfaces as shown in fig 32–2. The impedance of the throttle is given by the dimensions of the groove and is practically not disturbed by the admissible variations of the gap height. With very large gap heights, there will be a short circuit by a direct flow across the sealing lips.

2.  The throttle is built from a gap with a moving wall, as shown in fig 32–1 on the example between piston and cylinder. The essential fact is here that the gap height of the throttle, given diametric clearance, does not depend on the gap height of the slider or hydrostatic bearing it is feeding. The impedance can be calculated simply by the gap formula.

As a practical rule for throttling grooves, we can count on an approximate flow velocity of 60 to 80 m/sec with a pressure drop of 50–60 bar, corresponding to a dynamic pressure of 16–29 bar according to formula 13–2b.

For a more precise calculation we have to consider that the flow in the groove is usually near to the transition between laminar and turbulent conditions. The Reynold's number is, according to equation 13–4b, with $L = 0.1$ cm and $v = 60$ m/sec

$$Re = 2\,700$$

Under these conditions the pressure drop in a groove is built from the following contributions:

1.  Twice the dynamic pressure, referred to a continuous distri-
    bution of the velocity across the gap. This component is
    independent of temperature but increases with the square of
    the flow. The secant admittance decreases, therefore, with
    high oil pressure as is frequently desirable in order to limit
    the flow or leakage.

2.  The viscous pressure drop according to the gap formula, includ-
    ing the correction for a rectangular gap. It increases linearly
    with the flow, but is strongly dependent on the temperature
    because of the viscosity.

The main problem of throttling grooves is to provide sufficient
impedance, since possible wear of the sealing lips and the ease of
the manufacture require us to make the throttling grooves with a
width and a height of at least several hundred micrometres or tenth of
millimetres (or 10 milli-inches). This leads to much too large admittance in hydrostatic machines of the usual size.

The slipper in Section 25 requires a flow of 1.15 $cm^3$/sec and with a flow
velocity of 60 m/sec the following cross-section

$$A = \frac{Q}{v_m} = \frac{1.15 \ cm^3/sec}{6.10^3 \ cm/sec} = 1.9 \cdot 10^{-4} \ cm^2$$

corresponding to a dimension 0.14 x 0.14mm (6 x 6 milli-inches), a value too
small for easy manufacture.

The following methods are available for obtaining high impedances:

## 1. Intermittent oil injection

As shown schematically in fig 4, the throttling groove in the left
part is not running through to the depression, but terminates near to
it. The remaining sealing land is periodically short circuited by the
blind holes in the right part due to the relative movement. The blind
holes open the connection usually during 20% of the time, thereby
increasing the impedance by a factor of five. In addition, there will
be a small pressure loss due to the many accelerations of the oil
in the throttling groove. It is important to dimension the parts in
such a way that there is never a direct connection or short circuit
bypassing the throttling groove between supply bore and depression.

This intermittent oil injection has been known for many years and
works very satisfactorily in practice, especially in the valve faces
of axial piston machines. It can also be used for the axial shaft
bearings of tilting head machines, and for the slippers of swash

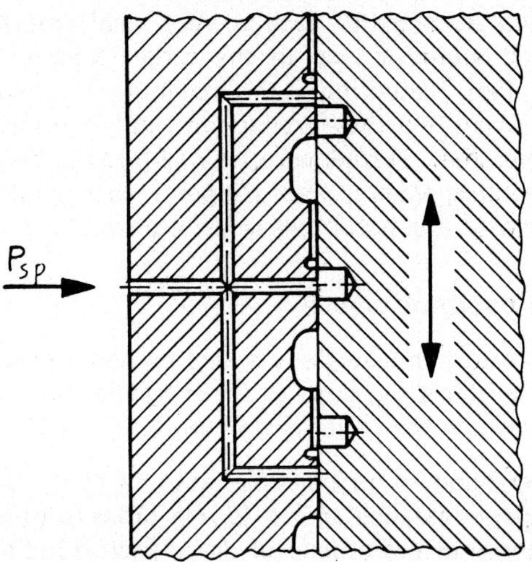

Fig 33.4: Principle of the throttle with intermittent oil
injection for increase of impedance

plate machines. Fig 5 represents an example of a throttling groove
on the spherical piston head, with a bore in the centre of the slipper
leading to an isle. This isle is short-circuited (or bridged) intermittently by the blind holes in the supporting surface at left, giving
an oil injection. In order to avoid a short circuit across the outer
sealing lips of the slippers, the diameter of the blind hole must be
less than the radial extension of the lips.

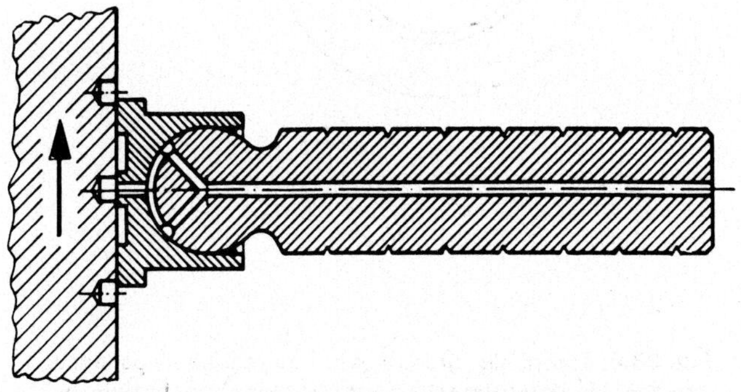

Fig 33.5: Slipper with intermittent oil injection

When working as a motor with a very small rotation frequency, irregularities of operation due to the intermittent oil injection are observed. For their reduction it is possible to dispose the blind holes in such a way that they produce several injections during each revolution of the shaft. Furthermore, we can arrange that the injections in the different components, like slippers, do not take place at the same time, but with a suitable phase difference.

### 2. Form of the throttling groove

It is possible to increase the viscous pressure drop by increasing the length of the throttling groove, for instance, by rolling it up like a spiral, as shown on the lower part of fig 6.

On the other hand, it is possible also to apply into a throttling groove (having for instance 0.5 to 0.5mm cross section) blind holes of 3mm diameter with a distance of 8mm. Each blind hole, therefore,

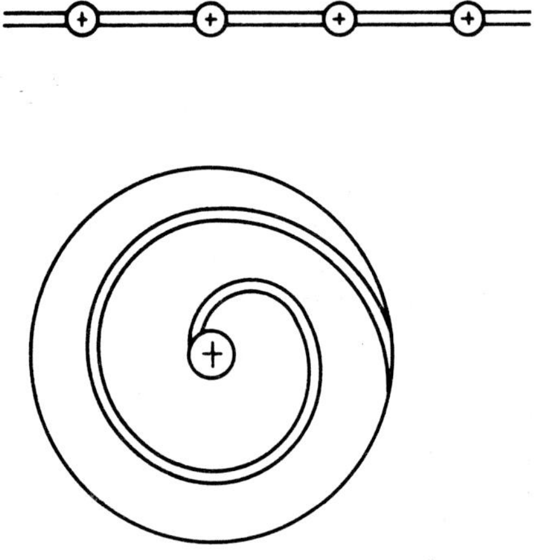

Fig. 33.6: Throttling groove with increased impedance, on top by enlargements (small holes), on bottom by increased length in the spiral

produces approximately a turbulent loss of twice the dynamic pressure independent of temperature. On the other hand, the (temperature dependent) viscous pressure drop is decreased since the active length is reduced by the blind holes. This method is represented schematically on the top of fig 6.

### 3. Metering by compression chambers

We can use the compressibility of the oil for throttling or metering by applying in the moving part a cavity, called compression chamber. This chamber is connected alternatively with the supply bore and with the depression, as shown schematically in fig 7. Unlike the intermittent oil injection, we do not have a short circuit, but the compression chamber with the volume $V_{cc}$ is filled first from the supply bore and discharges some oil when it registers with the depression at lower pressure. This volume is determined by compressibility modulus as follows :

$$\Delta V = V_{cc} \; \frac{\Delta P}{B}$$

and the flow of z compression chambers, discharging once each revolution:

$$Q = \frac{\omega}{2\pi} z \; V_{cc} \; \frac{\Delta P}{B} \tag{33–3}$$

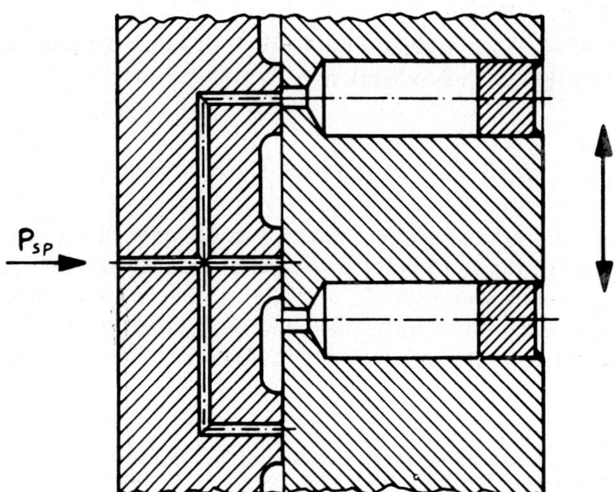


Fig. 33.7: Metering by compression chambers

We see from formula 3 that this is a linear throttling, where the flow increases proportionally to the difference of pressures in the supply bore and the depression. The admittance is, therefore, independent of the pressure, viscosity and temperature, but proportionately to the rotation frequency. /FN[1]/

Numerical examples : With $z = 9$ compression chambers having each $V_{cc} = 3cm^3$, we obtain as similarly in Section 25–1 :

$$Q = \frac{150 \text{ rd/sec}}{6.28}\, 9 \cdot 3 \text{ cm}^3 \, \frac{27 \text{ bar}}{1.5 \cdot 10^4 \text{ bar}} = 1 \cdot 15 \text{ cm}^3./\text{sec}$$

The difficulty with metering by compression chambers is the insufficient admittance in many cases, and to provide sufficient volume for them. Nevertheless, this method has great advantages for distributors and shaft suspensions in axial piston machines. It is less suitable for slippers, since the compression chamber also meters oil from the depression outwards. With high rotation frequencies the metering is very generous, but with low frequencies it is sometimes insufficient, and is subject to the same irregularities as intermittent oil injection.

Summarising throttling devices, there are three following throttling effects that can be combined in a device, depending on the desired characteristics and on the admissible complications:

1.    Viscous pressure drop,
2.    Pressure drop through inertia effect or turbulent pressure drop,
3.    Metering by compressibility

### 33–3   Geometry of the Cylinder Block

Consider the cross-section through the cylinder block of an axial piston machine, fig 8, in order to find the relation between the pitch radius $R_{zp}$ piston diameter $d_p$ and the number $z$ of cylinders. The most narrow part of the walls between the cylinders is situated on the chord between the axis of the cylinders. If the minimum wall

---

FN[1] — Strictly it would be necessary to include the expansion of the walls of the compression chamber. However, with thick walls, as for instance, with the compression chamber in the cylinder block of an axial piston machine, this correction is small. On the other hand, we can substantially increase the admittance with very thin walls. See also Section 72.

thickness is to be equal to the fraction $\delta$ of the piston diameter, then the length of the chord must be $(1 + \delta)\, d_p$

On the other hand, the length of the chord of a regular polygon can be read from fig 8:

$$(1 + \delta)\, d_p = 2\, R_{zl} \sin\phi \qquad\qquad \phi = \frac{2\pi}{2z}$$

Thus we obtain for the minimum pitch radius

$$R_{zl} = \frac{(1 + \delta)\, d_p}{2 \sin\phi} \qquad\qquad\qquad (33\text{--}4)$$

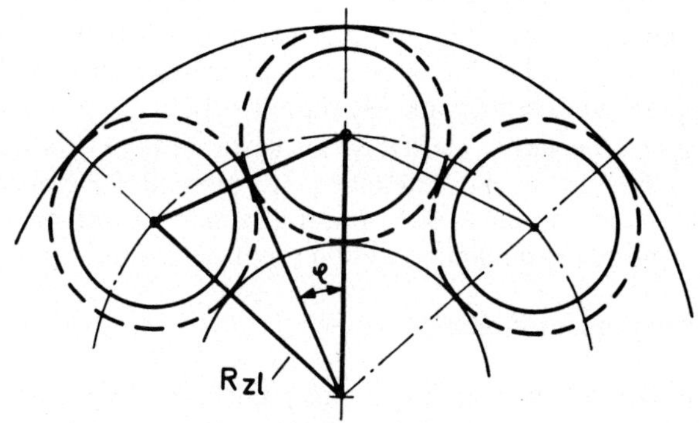

Fig 33.8: Schematic cross-section for the geometry of
the cylinder block

Ignoring the reinforcement by the outer part of the cylinder block, we can calculate the tensile stress in the narrowest part between adjacent cylinders with the formula of a thin wall tube :

$$\sigma = p\, \frac{d_p}{\delta\, d_p} = \frac{p}{\delta} \qquad\qquad\qquad (33\text{--}5)$$

where $\delta\, d_p$ = minimum distance between cylinders.

The minimum wall for tilting head machines oscillates between 25 and 33% of the piston diameter ($\delta = 0.25 - 0.33$), corresponding to a stress concentration /FN[2]/ of $4 - 3$ according to equation 5. It should be noted that higher stresses or thinner walls lead to a more variable design of the distributor.

---

FN[2] — Stress concentration means the ratio between tensile stress and operating pressure.

With swash plate machines, the length of the chord and therefore the pitch radius of the cylinder is imposed by the diameter of the slippers, in order to avoid fouling, giving sufficient wall thickness between adjacent cylinders.

The question of the optimum number of cylinders also belongs to the geometry of the cylinder block. Because of the much smaller fluctuations of delivery, usually only an odd number of cylinders is used /FN³/, and the choice is practically between 5, 7, 9 or 11 cylinders. Increasing their number within this range has the following advantages:

1. Delivery fluctuations decrease
2. The relative importance of the variations of the pressure balance of the valve faces becomes smaller, important especially for floating distributor (Section 33–5).
3. More space becomes available in the centre of the cylinder block inside of a wall thickness $\delta/2$ needed for strength. It can be used for universal joints (Section 33–6) or for the mushroom-formed balancing piston (Section 33–5).

and the following disadvantages:

4. The number of parts (pistons etc.) becomes larger and their size decreases, increasing manufacturing costs, especially in smaller machines.
5. With a given angle of inclination, the piston stroke becomes large compared to the diameter, increasing piston speeds and requiring slender connecting rods (in tilting head machines).
6. The total piston area that can be accommodated within a given diameter of cylinder block decreases.

Balancing the above requirements, the optimum seems to be with seven cylinders for tilting head and with nine cylinders with swash-plate machines.


## 33–4 Design of Valve Faces

Many axial piston machines use an axial valve face for the admission and delivery of the fluid to the cylinder block, collaborating with a valve face on the not rotating distributor or timing plate.

---

FN³ — Although one manufacturer has used ten cylinders for many years.

Fig 9 contains on the upper part a top view of the valve faces with the projection of the cylinder ports admitting the fluid from the cylinders to the distributing face in broken lines. These ports register ie establish a connection alternatively, due to the rotation of the cylinder block,with the kidney shaped slots, called kidney or timing ports, at right and left sides of the distributor, connected with the intake and the delivery of the machine. It is necessary to install the distributor in such a way, that the transition of the cylinder ports between both kidney ports correspond  to the dead centres of pistons movements. The contact faces around the ports constitute the sealing lips.

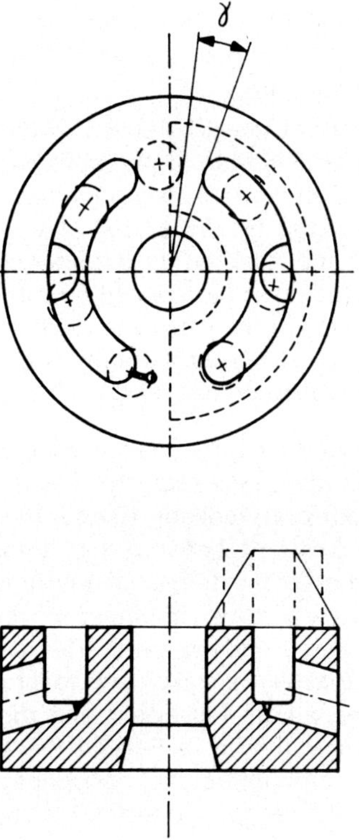

**Fig. 33.9: Distributor  with  kidney  ports  with  trans-
mission of the rest forces over the sealing lips**

The kidney ports terminate in the circumferential direction by a circle, corresponding to the form of a cylindrical milling cutter. Therefore, the angle during which the circular cylinder ports register

with the kidney ports is determined from the same geometric considerations as the chord length between the cylinders in Section 33–3.
Usually one chooses the circumferential extension in such a way
that there remains a lap angle, also called underlap or lap travel
/FN[4]/ of several degrees (usually 4 – 6°). The lap angle is denoted
with $\gamma$ on Fig.9.

The transition between high pressure and low pressure in the
cylinder takes place during the lap angle where the fluid is trapped
in the cylinder, corresponding to the steep parts of the working
diagram of fig 31–2. The pressure development depends on the
following effects:

1.  The remaining movement of the pistons, near the dead centres.

2.  The leakage due to gap flow across the sealing lips between
    cylinder and kidney ports.

3.  The compressibility of the oil trapped in the cylinders. A more
    gradual increase or decrease of pressure reduces noise produc-
    tion, and this is reason to make the lap angle as large as
    possible. It is limited by the piston movement that increases
    rapidly with distance from the dead centre,producing disturbing
    pressure peaks. They can be reduced by rotating the distributor
    through 2–3 degrees in the direction of rotation (in the case of
    a pump), leading to a correspondingly late switching from low
    to high pressure kidney ports. /FN[5]/

For further noise reduction larger lap angles can be used, up to
10 – 12°, but then it becomes necessary to compensate the increased
piston movement by artificial leakage flows. The throttling grooves
in the circumferential direction between the kidney ports as shown
in fig 9 serve this purpose. The length of the grooves is determined
by the condition that a lap travel of about 1° remains between the
end of the groove and the kidney port. The cross-section of the
throttling grooves is chosen in a way as to produce a continuous
pressure rise, under the piston movement and the variable pressure
drop, but the detailed calculation is beyond the scope of this book.
We only note that one calculates the compensation usually for an
operating pressure of 100 bar, $a = 1$ and for a rotation frequency of

---

FN[4] — This is the angle of rotation of the cylinder ports during which they do not
contact either kidney port.

FN[5] — This angular adjustment of the distributor in axial piston machines has a
similar significance as the change of the ignition timing in automobile engines.

$\omega$ = 160 rad/sec, and that throttling grooves of constant cross-section are more efficient than triangular grooves of increasing cross-section (vee grooves) as often used.

The calculation of the pressure field between the valve faces is important, since the corresponding forces have to be compensated by the piston forces acting in the inverse sense. If both valve faces are in parallel, we obtain a gap flow with linear pressure drop across the sealing lips, whilst over the depressions, that is the kidney ports and part of the cylinder ports, we always find the operating pressure. The extension of the pressure field in the circumferential direction varies with the rotation angle of the cylinder port, but in the mean it extends to the middle between the kidney ports, that is over a total angle of 180°. The lifting force with linear pressure drop is therefore given by:

$$F_a = P_{sp} \left[ A_{cv} + \frac{A_{lp}}{2} \right] \tag{33–6}$$

$A_{cv}$ = area of the kidney ports over 180°, with

$A_{lp}$ = area of the lips over 180°.

The valve faces are a kind of underbalanced hydrostatic bearing, as treated in Section 25–8. In many axial piston machines it is loaded by the forces of the pistons under pressure that push the cylinder block against the valve face. On average, one-half of the pistons is under pressure, but with an odd number of cylinders, the actual number fluctuates between the next entire numbers, that is between 3 and 4 in a 7 cylinder machine. This variation is compensated partly by a variation (of equal phase) of the pressure field between the valve faces, leaving a variation of the net force of about 5% in a 7 cylinder machine. /FN⁶/

In order to prevent the separation of valve faces, the balancing factor $\beta$ defined by equation 25–11 must remain well below one, especially in view of the above disturbances. This factor becomes with z pistons each with the area $A_p$

$$\beta = \frac{A_{cv} + A_{lp}/2}{A_p \, z/2} \tag{33–7}$$

---

FN⁶ — It is worthwhile to make a model of both valve faces in transparent paper and to study the pressure field on the valve face of the cylinder block depending on the rotative position of the distributor valve face.

Usual values of the balancing factor $\beta$ are 0.8 to 0.95, leaving a remaining force of 20% to 5% of the mean piston forces.

For reliable operation it is often desired to assure a determined distance between the valve faces, selected by the same criteria as the gap height of hydrostatic bearings in Section 25—7. Values of 10 to 15 micrometers usually give the most satisfactory results.

The balancing factor is determined by the selection of the areas of the kidney ports (as a depression extended over 180°) and of the sealing lips. The resulting areas are undesirably small, both for the cross-sections of the flow as for efficient sealing lips. The available area is distributed to lips and kidney ports in an optimum manner, leading usually to an area of the kidney ports of 64% and of the lips of 72% of the total area under pressure between the valve faces (corresponding to 54% and 62% of the piston area with $\beta = 0.85$). /FN[7]/

For an optimum sealing with a given lip area and gap height, the circumferential extension should be as small as possible (compare equation 25—5). This is obtained by reducing the pitch radius of the kidney ports compared to its width or compared to the pitch radius of the cylinder block, leading to more full or more radial extension of the lips. The question of reduced pitch radius is treated below in Section 33—5.

The pairing of materials on the valve faces should make no difference if a determined separation is maintained with full film lubrication. Nevertheless, a combination of tin or bronze with nitrided or case hardened steel has given the most satisfactory results. In order to avoid the manufacture from large blanks of bronze, some manufacturers apply thin layers of bronze on steel blanks, especially on the bores and valve faces of the blocks. The application is made by proprietary casting or welding processes, after which the cylinder block is machined to the desired final dimensions. Another solution with a steel cylinder block is to apply a thin bronze layer by electrolytic deposit on the steel pistons and on the valve faces of the cylinder block. One manufacturer claims that this coating is so accurate as to make subsequent machining unnecessary.

FN[7] — The pressure field on the valve faces is a half arc, and the pressure field of the pistons a number of circles lined up on the pitch circumference. The closer the cylinder bores are together, or the smaller the minimum wall thickness between them, the better, that is the more extended in the radial direction, become the kidney ports. This comes from the fact that the pressure field of the valve faces must also bridge the walls between the cylinders.

Instead of the balancing factor, Ernst (B3) uses a pressure ratio factor denoted as m as follows. The reciprocal of m indicates /FN[8]/ what the mean pressure on the lips would have to be in order to take up the remaining forces, instead of its value of 0.5 without such forces. This can be expressed in terms of the various areas as follows:

$$\frac{1}{m} = 0.5 + \frac{z}{2} \frac{A_p}{A_{lp}} [1 - \beta]$$

The values recommended above ($A_{lp} = 0.62\ A_p\ z/2$ and $1 - \beta = 0.15$) lead to $1/m = 0.74$ ($m = 1.35$) whilst Ernst recommends $1/m = 0.55 - 0.63$ ($m = 1.6 - 1.8$) (page 137 of Bibliography 3).

The remaining forces, not carried by the balancing factor of the kidney, must be transmitted from the cylinder block to the distributor by one of the following means:

1.  contact of the sealing lips
2.  hydrostatic bearings
3.  hydrodynamic bearings
4.  roller bearings

### 1. Transmission over the sealing lips

The remaining forces are simply taken by metallic contact of the sealing lips without special devices. Since the sealing lips must be continuous in the circumferential/direction, in order to avoid short circuiting the high pressure oil, the depressions mentioned in 26 cannot be used. This leads to unfavourable operation, but nevertheless such designs work in practice, but it is then recommended to use relatively small remaining forces, of the order of 5%. Any tendency to separation of the valve faces under extraordinary operating conditions can be prevented by a suitable abutment that limits the gap height say to 60 micrometres (2.4 milli-inches).

Valve faces with transmission of the remaining forces over the sealing lips are a kind of underbalanced hydrostatic bearing, as discussed in Section 25–8. They possess no hydrostatic stability since the pressure field is independent of the gap height, which is small and indeterminate /FN[9]/

---

FN[8] — The author fails to see why Ernst uses the reciprocal and not m directly.

FN[9] — Nevertheless gap heights of 20 micrometers have been measured and blamed on a geometric oil wedge with slightly inclined cylinder block, see E. Brangs : Axial-kolbenpumpen mit ebenem Steuerspiegel, Thesis TH Aachen, Germany 1965.

## 2. *Hydrostatic bearings*

Hydrostatic bearings are especially suitable for the transmission of the remaining forces since the lubricant can be taken from the delivery of the pump. In the example in fig 10 we find again in the centre kidney ports with sealing lips, limited by a circular groove called a draining groove, where the pressure is maintained equal to zero by suitable draining bores not represented. Outside there is the annular hydrostatic bearing with four depressions distributed evenly over the circumference.

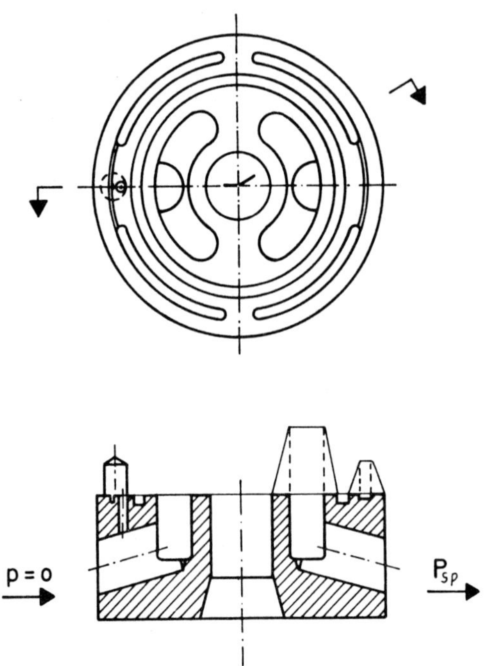


Fig 33.10:Distributor with hydrostatic bearing and intermittent oil injection by blind holes in the cylinder block (left) and by small throttling groove and pressure distribution (right above)

Each of the admission and delivery ports of the machine feeds two of the depressions of the hydrostatic bearing over a suitable impedance. Therefore, with normal operation only the two depressions near to the high pressure side of the machine are under pressure. This correlates well with the fact that the remaining forces attack eccentrically on the high pressure side.

The throttle can have a variety of forms, but in fig 10 it is a groove in the circumferential direction between the two depressions. The high-pressure oil is conducted from the delivery ports (or from the admission port with inverse sense of pressure) by a small bore between the two depressions on a point of the sealing lips in the proximity of the throttle groove. The blind holes in the cylinder block establish an intermittent connection between these bores and the throttle groove, leading to intermittent oil injection. The calculation of the impedance has been treated already in Section 33—2, but as a practical rule, the throttling groove has a depth and width of 2‰ (0.2%) of the piston diameter.

It is quite possible to use different throttling devices. Considerably increasing the size of the blind holes without the throttling groove they act as compression chambers metering the oil from the feeding bores to the depressions as described in Section 32-2. Furthermore, both methods can be combined by using a very fine throttling groove admitting oil with low rotation frequency, whilst the large blind holes act as compression chambers admitting more oil with higher rotation frequencies. In a sense we have then two throttles in parallel each with its own characteristics as described in Section 32—2, and the total admittance is equal to the sum of their admittances.

Fig 11 represents a similar design with a continuous annular throttling groove and a large number of shallow holes, instead of the depressions. The intermittent oil injection is made as before by feeding bores in the middle and by blind holes in the cylinder block that make an intermittent connection to the circular groove. Furthermore, the figure represents the spherical distributing surfaces which we shall further discuss in Section 32—5.

A further method is to take the oil in the gap between the pistons and the cylinders by suitable bores, as shown in fig 32—1. This is then throttling in a gap with moving walls, disturbed a little by the piston movements.

In order to increase the admittance with narrow piston clearance, it is advisable to provide suitable collecting grooves in the piston or in the cylinder wall.

Fig. 33.11: View of a frequently used distributor with
hydrostatic bearing similar to Fig.10 with annular
throttling groove

### 3. Hydrodynamic bearings

We obtain a hydrodynamic bearing by arranging a number of pads
(or isles) with radial draining grooves around the sealing lips outside
a circular draining groove. With a slight inclination in the direction of
rotation as shown in the former fig 26—1, there will be a pressure
field due to the oil wedge effect that can give the desired oil film on
the sealing lips. Such hydrodynamic bearings are frequently used
with pumps running always in the same direction and with high
rotation frequency. They do not contribute to the leakage flow of the
machine since no oil is taken from the delivery ports, but their
disadvantage is the insufficient pressure development at slow sliding
velocities. Furthermore, they are effective only in one direction of
rotation. This could be eliminated by giving to the isles an inclination
on both sides, but this reduces the pressure development at given
gap height and thus the admissible load. /FN[10]/

FN[10] — It is also possible to combine hydrostatic with hydrodynamic bearings, the
latter giving an additional lift at high sliding velocities. This could be done by
introducing throttled high pressure oil in the pads of fig 26—1.

### 4. Roller bearings

Some designs provide an accurately manufactured step instead of the external ring that accommodates rollers or needles in a suitable cage. They work either directly on the cylinder block, made from ball bearing steel, or on an inserted steel ring with cylinder blocks in bronze. The distance between the valve faces is determined by the diameter of the rollers, together with the depth of the step.

While it does not seem to be extraordinarily difficult to obtain the necessary accuracy (around ± 3 micrometres or ± 120 microinches) for the depth of the step, the main trouble is that the rollers wear some micrometres over several hundreds of hours. The wear reduces the gap height and increases the danger of seizing, but one manufacturer avoids this danger by applying on the sealing lips a layer of antifriction plastic containing PTFE.

## 33–5   Suspension and Force Compensation of Cylinder Blocks

For the proper operation of axial piston machines, it is essential that the distributing surfaces run with a small separation and parallel, as is best obtained by self-alignment. For this purpose, the cylinder block is suspended so that it can make small adjusting movements. The different forces are compensated in such a way that the valve faces come in the desired contact under the effect of the small remaining forces.

With axial piston machines we obtain compensation of the piston forces and of the forces of the axial valve faces, as described in Section 33–4. Apart from the forces, it is necessary to compensate also the various tilting moments, in order to obtain a satisfactory contact, as we shall see below.

### 1. Simple suspension

An arrangement of axial valve faces with compensation of the forces, except small remaining forces, is represented in fig 12. The cylinder block is guided with a suitable clearance /FN[11]/ on a central spindle suspended both in the drive flange and in the distributor. The spherical member in the drive flange and the clearances on the contact areas of the central spindle in the cylinder block and the distributor allow slight adjusting movements. This means that the cylinder block can rotate around both transverse axes until the valve

FN[11] — Instead of the clearance, spherical surfaces could be used.

faces of the cylinder block and the distributor are exactly parallel. The kidney ports are on the same pitch radius as the cylinder bores, thereby avoiding any tilting moment. The vectors of the different forces are represented by small arrows in fig 12.

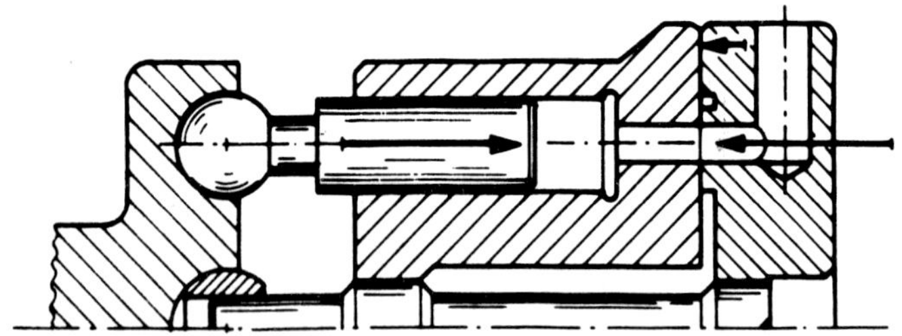

**Fig. 33.12: Suspension of the cylinder block with kidney ports on the pitch radius of the cylinder**

Such a suspension with the greatest liberty of movement or the minimum number of constraints is called kinematic design /FN[12]/ It is based on the fact that each object has six degrees of freedom in space,and the principle is to constrain or to determine each co-ordinate only once, so that hyperstatic conditions are avoided. In the case of the cylinder block, the contact with the central spindle determines two co-ordinates, namely the positions across the spindle, allowing small rotations around both lateral axes. /FN[13]/ The contact of the valve faces determines this angular position, fixing thereby two more co-ordinates. At the same time it determines one more co-ordinate, ie the position of the cylinder block along the axis of the spindle. Therefore, only one degree of liberty remains, the angular position around the axis, needed for the rotation of the cylinder block. Kinematic design is most useful not only in hydrostatic machinery, but also for instruments of physics and other parts of mechanical engineering.

## 2. *Spherical valve faces*

The disadvantage of the arrangement of the kidney ports on the pitch circle of the piston is that they become relatively narrow or small in the radial direction, due to their limited area. The available area, equal to a fraction of $1 - \beta$ (about 80 to 95%) of the area of half the pistons is distributed on the sealing lips and on the kidney ports.

On the other hand, the circumference of the sealing lips should be as small as possible, in order to reduce the leakage (Section 25).

FN[12] — Known since J.C.Maxwell in the 19th century for instruments in physics.

FN[13] — These rotations are unlimited if a spherical member is arranged between central spindle and cylinder block.

This can be obtained by reducing the pitch radius of the kidney ports, compared to the pitch radius of the cylinders. However, this produces a tilting moment tending to tip the cylinder block in the plane of the paper to the right, since the vector representing the centre of the pressure field around the kidney ports is offset towards the axis.

One way to avoid the difficulties of the tilting moment is to make the cylinder block adjustable only lengthwise, but to prevent any rotation around transverse axes by suspension in suitable needle bearings, which are then loaded by the tilting moment. This is no longer a kinematic suspension of the cylinder block, since the angular position of the valve faces depends on the accuracy of the seats and on the clearance of the needle bearings.

The spherical distributor in fig 13 allows us to reduce the pitch radius of the kidney ports, compensating the tilting moment with full kinematic adjustability. Due to the spherical form of the valve faces, the vector of the pressure field, always perpendicular to their centre, is slightly inclined outwards and transmits a radial force component to the cylinder block. We choose the radius of the spherical valve faces so that this vector of the pressure field intersects the axis of the cylinders at the same distance as the cylinder block is guided on the central spindle, as shown in the figure. The guiding allows again small adjusting rotations around the lateral axis and a contact of the valve faces without tilting moment. The radial force, introduced by the valve faces, is conducted out from the cylinder block on the central spindle at the guiding area. The vector of the pressure field exerts no torque on the point of inter-section and therefore no tilting moment is generated in the cylinder block. Any remaining torque, such as is due to the various tolerances, is transmitted with the remaining force to the valve faces.

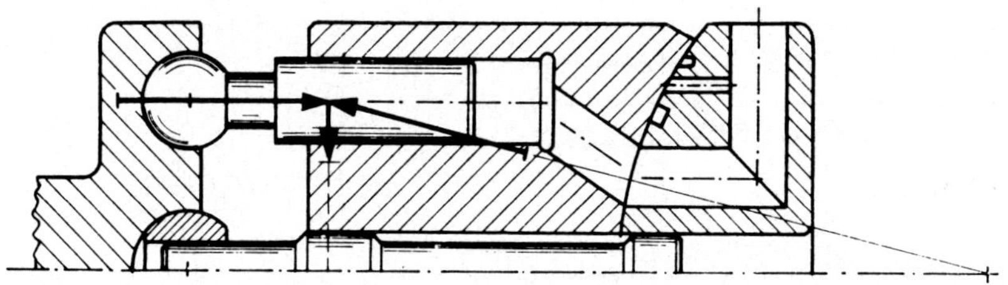

Fig 33- 13: Suspension of the cylinder block with kidney ports on reduced pitch radius and spherical valve faces for compensation of the tilting moment

Another method of describing the situation is that a second tilting moment is applied on the cylinder block by the radial component on the pressure vector of the valve faces and by the corresponding radial force of the guiding area further to the left, compensating the tilting moment due to the reduced pitch circle of the kidney ports.

The suspension with spherical valve faces has many advantages, and their only disadvantage is the slightly difficult manufacture of the spherical faces with high accuracy, for which many procedures are known from lens manufacture in optics. Furthermore, the central spindle transmitting the radial force component is subject to the corresponding bending moment.

### 3. Balancing piston

Fig 14 shows another design with plane valve faces and reduced pitch radius of the kidney ports. Here the central spindle is prolongated to the right and ends in a balancing piston free to rotate in an expansion cylinder. The non-rotating expansion cylinder abuts over a small spherical face on the rear of the distributor and is supplied by the bores with non-throttled, high-pressure oil from the kidney ports. Therefore, the piston generates a force to the right which is transmitted centrally by the central spindle on the cylinder block, as represented in fig 14 by the vector on the axis of the spindle. Its magnitude is selected so that it compensates, together with the vector of piston areas, the vector of the forces of the kidney ports for magnitude and position (or moment). This allows us not only to reduce the pitch circle but also to increase the area of the kidney ports and sealing lips, since now the area of the expansion cylinder is also available. Usually a balancing piston having an area of 40 to 50% of the area of the pistons under pressure is chosen.

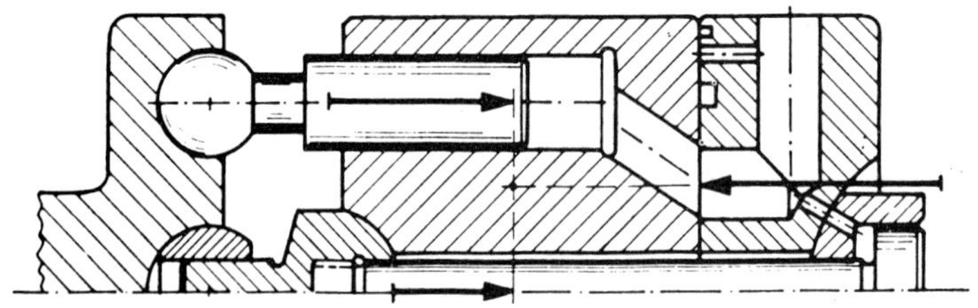

Fig. 33.14: Suspension of the cylinder block with kidney
ports on reduced pitch radius and with expansion
cylinder on the axis

With hydrostatic machines required to produce pressure in both directions, the expansion cylinder can be supplied from both kidney ports over non-return valves, or it is possible to use two separated concentric expansion cylinders.

Other designs use expansion cylinders in the cylinder block between the working cylinders without a central spindle going through In its place there is a pivot in the form of a mushroom coming out from the distributor and ending in a piston, collaborating with an expansion cylinder in the cylinder block.

### 4. Floating distributor

Fig 15 represents a cylinder block with reduced pitch radius of the kidney ports on the valve face, suspended rigidly on the central spindle by two separate guiding areas. Therefore, it is not angularly adjustable and normally is also fixed in the longitudinal direction. Instead, the distributor is suspended, slightly rotatable, about both transverse axes and is also adjustable in its longitudinal position. For this purpose there are several sleeves disposed in bores in the distributor and in a fixed part of the casing further to the right. The sleeves allow slight adjusting movements and press the distributor against the cylinder block according to their area, transmitting the oil at the same time between distributor and casing.

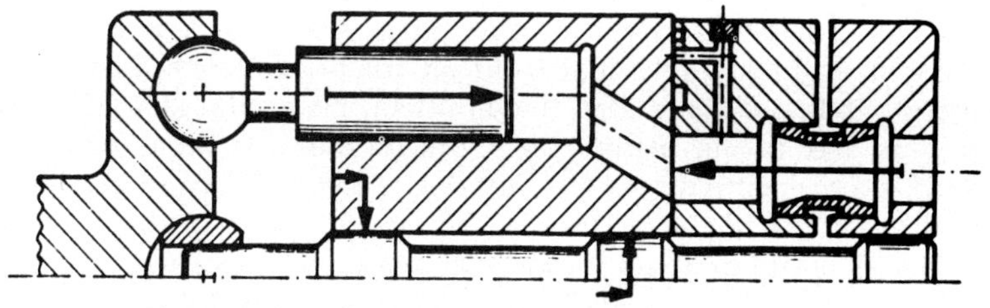

Fig 33.15:Cylinder block rigidly suspended on central
spindle with kidney ports on reduced pitch radius and
with floating distributor

Whilst the tilting moment on the cylinder block is now taken up by the rigid connections with the central spindle as shown, it is necessary to choose for the sleeves such dimensions and positions that the force of the valve face of the distributor is compensated in its magnitude and also in its tilting moments. Normally one uses two sleeves for each kidney port, offset from the plane of the paper of fig 15. Sometimes additional blind pistons in suitable positions are necessary.

In selecting the dimensions of the sleeve we must keep in mind that we now need a remaining force pressing the distributor securely against the cylinder block. Since the pressure field around the kidney ports fluctuates according to the number of pistons under pressure, and since the force of the sleeve is constant, it is necessary to provide more remaining force than with floating cylinder blocks. Depending on the geometry of the kidney ports one selects a cross-section of the sleeve 120% to 125% of the mean area of the valve faces under pressure.

This method of suspension with fixed cylinder block and floating distributor has advantages, especially with swash plate machines. This comes from the fact that in this type the lateral forces of the pistons act on the cylinder block and must be transmitted to the shaft, disturbing the exact compensation of forces necessary for self-adjustment of cylinder block. Nevertheless, such a design has also been used in practice.

The disadvantage of the floating distributor is that it imposes more load on the bearing between the valve faces due to the force variations, and that it needs a little more length, as can be seen in fig 15.

### 33–6    Synchronisation between Drive Flange and Cylinder Block

The radial forces and the useful torque are produced in tilting head machines in the spherical seats of the drive flange, since the connecting rods and the piston forces attack there obliquely, corresponding to the tilting angle. The cylinder block is free of radial forces if the connecting rods are parallel to the axis of the cylinder, because the force vector of the connecting rods with its ball joints is always along its axis, as known from elementary mechanics.

It is important to prevent any rotation between cylinder block and drive flange, or in other words, their rotating angle must be well synchronised. Any error produces an inclination of the connecting rods in the circumferential direction. This generates tangential forces and a torque on the cylinder block that tends to increase the angular misalignment. The tangential inclination of the connecting rods is given by

$$\gamma_{cr} = \frac{R_{fl}}{L_{cr}} \, \Delta\phi \qquad\qquad (33\text{–}7)$$

with $\gamma_{cr}$ = tangential angle of the connecting rods

$R_{fl}$ =   pitch radius of spherical seat in the drive flange

$L_{cr}$ =   length of connecting rod between the centres of the spheres

$\Delta\phi$ =   angular error of synchronisation

Another way of stating this is that the connecting rods loaded by the piston force are in an unstable equilibrium with their starting inclination. Any increase of the relative rotation must be prevented by the synchronisation because it would otherwise increase the loading torque.

For the synchronisation between drive flange and cylinder blocks the following methods are known:

## 1. *Universal joint*

A universal joint (also called Cardan joint) between drive flange and central spindle is connected to the cylinder block in such a manner as to prevent relative rotation around its axis, but to allow adjusting movements around both transversal axes.Since conventional universal joints have irregularities of rotation increasing with the tilting angle, it is necessary either to use a double universal joint or one of the newer homokinetic (constant velocity) universal joints.

The disadvantage of this design is the cost and the fragility of the universal joint and also that the synchronisation or angular alignment must be adjusted carefully during assembly. Any play or beginning wear allows a tangential inclination of the connecting rods, increasing the torque on the universal joint and probably also the wear. On the other hand, it has the advantage that large tilting angles are possible, up to 35° in industrial practice.

## 2. *Synchronisation by connecting rods*

A modern method is the synchronisation by contact between the piston skirt and the connecting rods, as represented in fig 32–4.

In order to study the geometry, fig 16 represents the pitch circle of the cylinder bores. As we see from there, the pitch circle of the spherical seat in the drive flange appears as an ellipsola depending on the tilting angle, with the half axis.

$R_{fl}$   and $R_{fl} \cos \hat{a}$

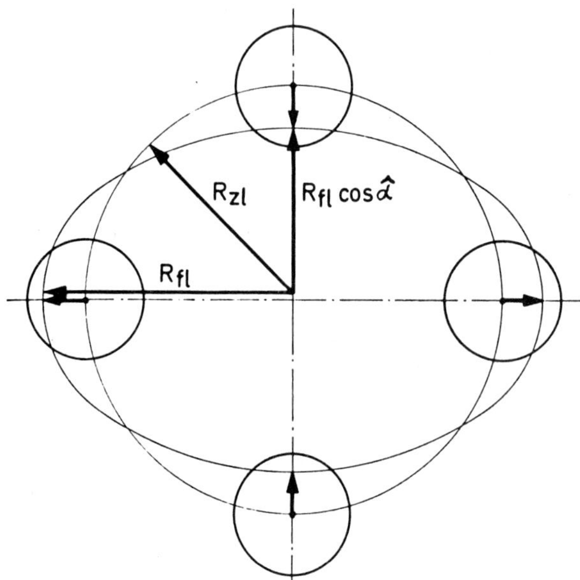


Fig 33.16: Geometry of synchronisation by connecting rods

The inclination of the connecting rods with respect to the cylinder axis depends on the difference between the ellipsola and the pitch circle of the cylinder bores. In order to reduce these deviations, the pitch radius in the flange is chosen a little larger, so that the pitch radius of the cylinder is situated in the middle between the large and the small half axes of the ellipsola. With a maximum tilting angle of 25° this leads (because of cos $\hat{a}$ = 0.91) to a 5% increase of the pitch radius in the drive flange compared to the one in the cylinder block.

The movement of the connecting rods is represented by small vectors in fig 16. It is advisable to make the angular liberty of the connecting rods before abutment on the piston skirt a little larger in order to allow for errors and tolerances. Usually there remains liberty for a relative rotation between drive flange and cylinder block of 1° in small and 0.5° in large machines. The corresponding angular liberty of the connecting rods in the piston skirt depends on the length of the connecting rods and can be calculated by equation 7.

The synchronisation by connecting rods is very reliable, since the remaining torques are transmitted on a large radius. Its disadvantage is a certain sensitivity against misalignment between the cylinder block and the drive flange because of its effect on the angular position of the connecting rods. Therefore, a good suspension of the cylinder block, as discussed in Section 33–3 is important here. The possible maximum tilting angle is limited to about 28°, since the diameter of the pistons must be allocated to the diameter of the connecting rod, to the wall thickness of the piston skirt and the angular liberty of the connecting rods.

### Bevel Gears

Bevel gears produce a very reliable and robust synchronisation. They can be applied either on the central pivot and on the centre of the drive flange, or outside around the cylinder block and the drive flange, an example of which has been presented in fig 32–7. The great limitation is that bevel gears can be built only for a determined angle, and cannot, therefore, be used for hydrostatic machines with adjustable displacement.

## 34   Hydrostatic Machines without Pistons

Apart from the hydrostatic piston machines treated in Section 33, there are many other types of hydrostatic machines used in industry. They can be classified as follows:

1. Gear machines
2. Vane machines
3. Other hydrostatic machines

These displacement machines without pistons are called sometimes gear pumps or vane pumps, but such a limitation is not appropriate, since there exist both gear motors and vane motors.

In piston machines the displacement chamber is built by the piston and the cylinders, which can relatively easily be manufactured to the required accuracy. In other hydrostatic machines the sealing depends on the simultaneous accuracy of many dimensions. Consequently leakages are generally higher and such machines are less suitable for high pressure.

In theory  there are many mechanisms that can be used for the displacement of fluid, and only a small number have reached industrial importance. For a general view of possible displacement mechanisms the reader is referred to the excellent book of Hadeckel (B8) but the old book of Reuleaux (B12) is also still worthwhile reading.

Basic for the design of displacement mechanisms and particularly for sealing against fluid pressure are the following two kinds of contact between bodies or component parts:

1. Line contact
2. Area contact

With line contact, the bodies are in contact along a line, in principle infinitely thin as long as the local deformations of the contact points can be ignored. Examples are the cylindrical roller on a plane and spherical piston in a cylinder. The sealing effect and the elastic stress depends according to the Hertzian formulae on the radius of the bodies in contact, as we have seen in Section 26–2.

With area contact the bodies are touching each other over a surface or area, like a cylindrical piston in the bores of a cylinder. The contact zone has, therefore, a certain extension, and any leakage must flow through a gap with the corresponding length. Thus our gap formula becomes applicable and one obtains a useful sealing effect with gap heights of several micrometers. Opposed to this we have seen, for instance with spherical pistons in Section 23, that the leakage across a line contact (even with large radius) is much larger.

## 34—1    Gear Machines

A gear fitted tightly into a casing entrains fluid in the circum‑ferential direction in the spaces between the teeth. In order to obtain a useful delivery, one uses two meshing gears where the liquid is displaced during the penetration of the space between the gears and forced into the delivery ports.

Let us consider fig 1 for the operating principle. It shows the development of two gears or rather of four racks, of which both outer ones travel to the top, and both inner ones to the bottom of the figure. The internal racks are meshing and prevent by means of the contact points (drawn as small circles) a return of the liquid.

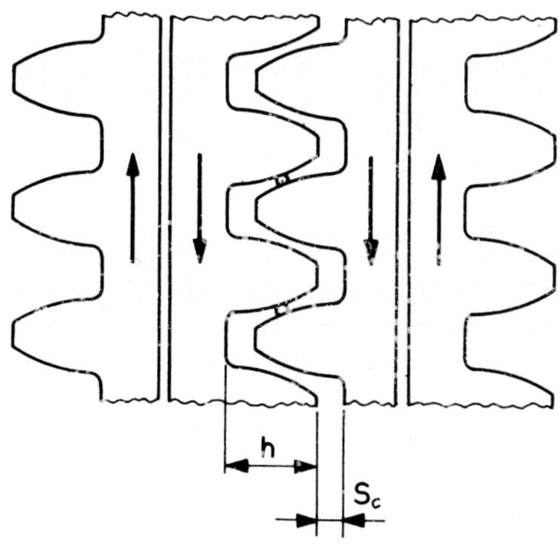


**Fig 34.1: Principle of the gear pump represented by four racks**

Each space between the teeth of both outer racks delivers the following volume to above

$$q' = 2\,[L_p\,h - A_t]\,b$$

with $L_p$ = circular pitch

$\quad h\ \ $ = tooth height

$\quad A_t$ = cross-section of a tooth

$\quad b\ \ $ = width of the tooth

The meshing inner racks transport a volume backward determined by the pitch, and the distance between the bottoms of the spaces between the teeth (dedendum circles) on both racks. From this we have to deduct the cross-section of both meshing teeth, one coming from the right and from the left rack.

$$q'' = 2\,[L_t\,(h + s_c) - 2\,A_t]\,b$$

with $s_c$ = tooth tip clearance (between the dedendum and addendum circles).

The resulting displacement volume per tooth is the difference of the tooth volume

$$q-q'' = L_t\,(h - s_c)\,b$$

and the displacement volume per radiant of a machine with z teeth per wheel becomes

$$q = \frac{z\,L_t}{2\pi}\,(h - s_c)\,b \qquad\qquad (34\text{–}1)$$

We obtain therefore the remarkable result that the displacement is independent of the cross-section of the teeth. This comes from the fact that with more full cross-section the delivery on the outside diminishes, but at the same time also the reverse delivery through the meshing zone, so that the net delivery remains unaffected.

Using the diametric pitch $m = L_t/\pi$ we obtain

$$q = \frac{z\,m}{2\pi}\cdot(h - s_c)\,b \qquad\qquad (34\text{–}2)$$

In standard gears we have $h = 2.2\ m$ and $s_c = 0.2\ m$, giving a displacement per radiant of

$$q = z\, m^2 b = d_t\, m\, b \tag{34–3}$$

with $d_t = zm$ = diameter of pitch circle. This result remains valid if we replace the racks by two meshing gears fitted in a casing as shown in fig 2.

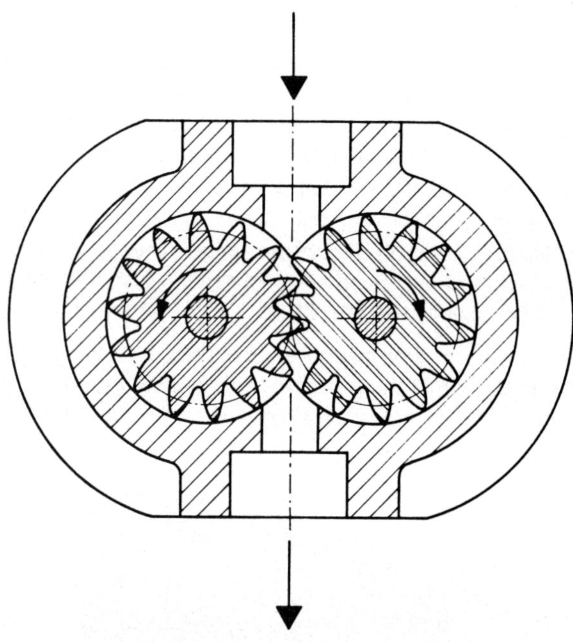

Fig 34.2: Schematic   cross-section   of   a   gear   pump

Essential for operation is the sealing by the meshing or contact points of the gears. The instantaneous delivery flow depends on the length of the radius vectors between the contact point and both axes, variable with rotation angle and producing fluctuating flow. We do not treat this problem further here, but we would like to mention that normal gears have frequently two meshing points, ie two points of contact between different teeth in fig 1 and 2. This leads to two different flows, practically to trapping, meaning compression or cavitation of the oil trapped between both meshing points. In order to avoid pressure peaks suitable depressions or grooves can be applied in the side covers of the casing. They bridge or short-circuit one contact point so that only the other one remains for sealing.

With pressure exceeding 10 to 20 bar (150 to 300 psi) the leakage becomes important, composed of the contributions of the following leakage flow path :

1.  Leakage flow through the contact (meshing) points
2.  Leakage between gear tips and the stator ring
3.  Leakage along the side faces

Whilst the leakage through the gear mesh is usually small due to the contact pressure under the transmitted torque, the other leakage contributions depend on the accuracy of manufacture. The leakage over side faces is especially troublesome, but it can be considerably reduced by having slightly movable side walls loaded by the delivery pressure. For this purpose one introduces into the casing a displaceable bushing or gear cover, one face contacting the gear whilst delivery pressure acts on the other face. The pressure fields are limited by sealing rings in order to obtain a suitable force pressing the cover and the gears together.

Fig 3 represents a gear machine with axially moveable gear covers for taking up the clearances. The figure also shows very well the general arrangement of a practical gear machine.Another manufacturer

**Fig 34.3: Gear pump with pressure-loaded side faces or side covers (Hydroméca)**

also includes a radial compensation for taking up a clearance between gear tips and stator rings. This is done by equipping the gear covers with a further pressure field limited by a sealing ring on the circumference, exerting radial forces pressing the gear tips against the stator rings.

Other forms of gears may be used. Fig 4 represents a remarkable annular gear pump, with axial and radial compensation of clearance. The space between the gears is divided into two parts by the mesh points and by the finger-like filler piece 4. Only the smaller part space or chamber is filled by delivery pressure. The oil enters the large chamber through the bores in the annular gear. From there it is delivered in the spaces between the teeth into the smaller chamber, as sealed by the filler piece 4, and thereby pressurised. Finally, it is expelled before the mesh point through the bores of the annular gear into a bridge piece 6 loaded radially by the pistons 7. The lateral gear cover 5 compensates the axial clearances by means of the pressure fields 8. Due to the long mesh of the teeth the operation of the pump produces little noise, and the manufacturer claims it suitable for 240 bar (3800 psi) in continuous operation.

Hydrostatic gear machines of various designs are very widely used. Normally they are limited to pressures of 140 bar (2000 psi),

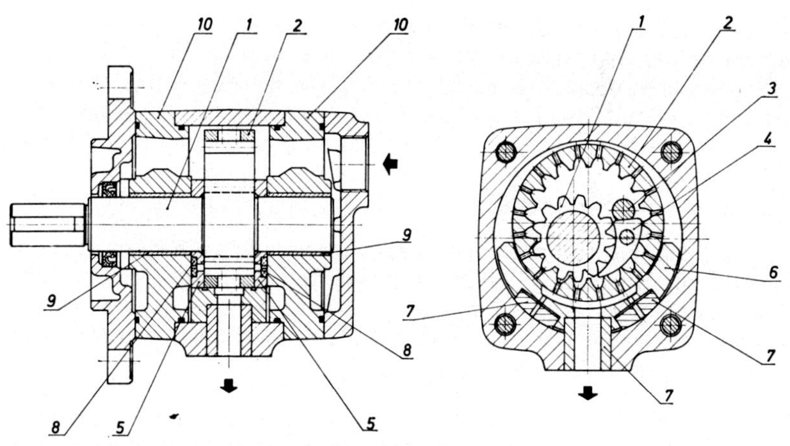

**Fig 34.4: Cross-section of an annular gear pump with axial and radial compensation of clearance (Eckerle)**

and to displacement volumes of about $100\,\mathrm{ccm^3/rev}$. However, in recent years new designs have appeared claiming higher perform- ances.

The pressure chamber in gear machines produces a considerable radial force on the gears which can be taken up either by needle bearings or by bearing bushes, sometimes covered with an anti- friction (PTFE) layer. Unfortunately, it is practically impossible to build gear machines with adjustable displacement volume.

It would be possible, in principle, to build gear machines with adjustable displacement volumes, and such a design was once produced in the USA. There one of the gears could be axially displaced partly out of mesh and replaced by a non-rotating filler piece. Depending on this axial displace- ment of gear and filler piece the effective width and, therefore, the displace- ment according to equation 3 was varied. The main difficulty here was the sealing of the many gaps between the pieces.

## 34–2   Vane Machines

Vane machines are another frequently used design. They consist of a rotor revolving in a track ring in a case, equipped with radially adjustable vanes for entraining the fluid in the circumferential direc- tion.

The most simple design of vane pump uses a circular rotor disposed eccentrically in a circular track ring. As shown in fig 5, the rotor is equipped with slits accommodating radially movable vanes of exactly the same width. The displacement chambers are located between the rotor and the track ring and are limited in the circumferential direction by the vanes. They change their volume periodically due to the eccentricity of the rotor and can   thereby be used for delivery.

As with piston machines, each compression chamber must be connected   alternately  with the intake and with the delivery ports. The simple method is shown in fig 5 with suitable milled ports or slits in the track ring. More usually, one uses instead ports in the lateral covers, as we shall see in fig 6.

As with gear machines, the delivery consists of the flow trans- ported between the rotor and the casing from the intake to the delivery port, on top of fig 5, from which the reverse delivery between the rotor and the bottom of the casing is deducted. This allows us to establish an approximate formula for the displacement volume. With

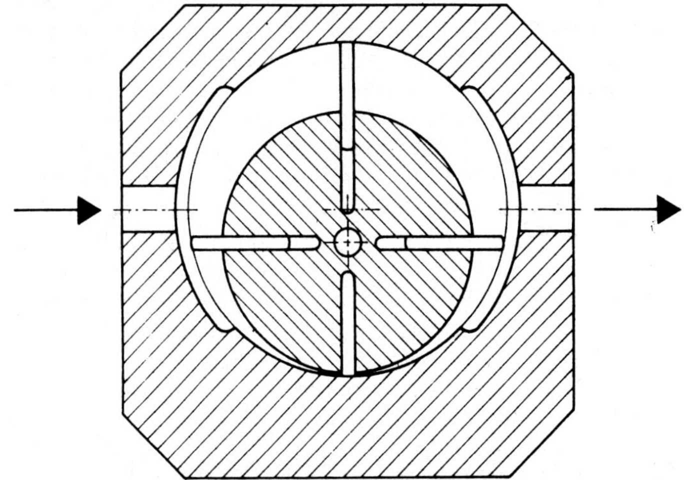

Fig 34.5: Schematic cross-section through a vane pump

an eccentricity e of the rotor, the maximum height of the displacement chamber on top is equal to 2e and on the bottom equal to zero if
the rotor just touches the track ring. The mean circumferential velocity
is equal to the radius of the centre of the displacement chamber R + e
multiplied by the rotation frequency. With a width b perpendicular to
the plane of the paper we obtain for the flow

$$Q = \omega \, (R + e) \, 2eb \tag{34-4}$$

and for the displacement of a vane machine

$$q = 2(R + e) \, eb \tag{34-5}$$

The displacement is unchanged if the rotor with the same eccentricity is fitted into a larger track ring, because then both the forward
and the reverse delivery are increased by the same amount.

Equation 5 is strictly valid for a vane machine having a large
number of very thin vanes. In practice one uses a small number of
vanes with non-negligible thickness or extension in the circumferential direction, with the following consequences:

1. The displacement volume is reduced by the volume of the vane
   parts projecting outside the rotor, unless the space behind the
   vanes in the slits is also used for delivery.

2. The flow is now composed of the deliveries of the single displacement chambers. As with piston machines, this produces delivery variations, smaller the greater the number of displacement chambers or vanes. Furthermore, the delivery is reduced a little since it no longer depends on the maximum height between rotor and starter ring, but only on the maximum mean height between two vanes.

3. The ports for admission and delivery of the fluid are designed as in axial piston machines. Their circumferential extension must be limited so as not to produce a short-circuit in any position of the rotor, but rather give a few degrees of lap travel similarly as discussed in Section 33–4. This means that the land between the distributing ports is a little more extended than the distance between the vanes.

Vane machines, according to fig 5, can be built with variable displacement by adjusting the eccentricity of the rotor in the stator or track ring, and are actually manufactured in large numbers.

The rotor of fig 5 experiences a radial force in the direction of the intake port, given by the product of the width of the rotor, the diameter of the stator track ring and the delivery pressure. This force must be taken up by the shaft bearings, limiting the operating pressure of simple vane machines to about 100 bar (1500 psi).

Many designs of vane machines have two pressure zones arranged diametrically opposite to balance the radial forces on the rotor. This needs an eliptical stator ring in the casing and two each of admission and delivery ports, but otherwise is designed according to the same principles. The vanes have then two movements per revolution, thereby doubling the displacement volume. The depressions in the rotor between the vanes allow us to provide larger cross-sections for the distribution ports as can be seen in fig 6, but they do not influence the delivery.

The displacement of a double vane machine can be determined again approximately by formula 5, multiplying by the factor 2. Fig 6 represents the practical design of a double vane machine, produced in the range of 0.95 to 6.4 cm$^3$ per radiant (= 2.3 up to 40 cm$^3$/rev). The casing carries a stator ring with elliptical track and the rotor is suspended by the bearings in the covers at right and left. The intake is on the right-hand side and the delivery takes place through the pipe connection at the left.

A special feature of fig 6 is the *clearance compensation* of the side faces. This is obtained by an axial movement of the left side cover, the rear side of which is partially subject to delivery pressure producing a corresponding force. It is seen that this vane machine has only few parts requiring high accuracy in manufacture. These parts have furthermore a very simple form, whilst the more complex form of the casing does not need to be produced accurately.

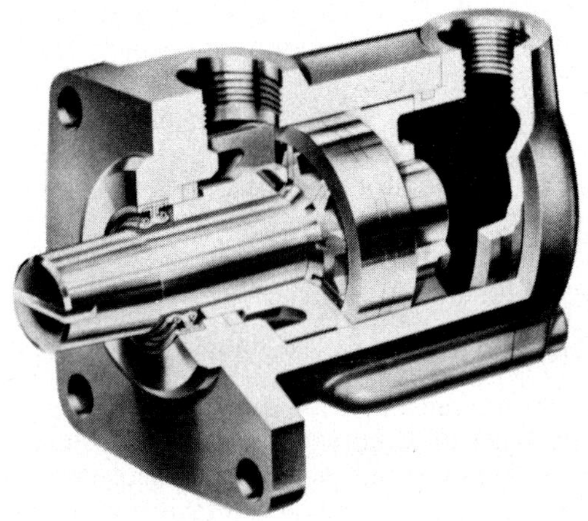

Fig 34.6: Cross-section through a vane pump with compensation of axial clearance (ATE)

The great disadvantage of the compensation of the radial forces by two pressure zones in the elliptical stator is that such machines cannot be made with variable eccentricity and therefore not with adjustable displacement. As an exception there exists a new development with an elastically deformable stator ring, where the form of the track of the vane and the displacement volume can be adjusted by small cylinders, attacking the stator ring from the outside . /FN[1]/ Any distortions of the track of the vanes with consequent liability to seizure remains an open question.

Due to the mobility of the vanes, no great accuracy of the track diameter nor compensation of the radial clearances (as in gear machines) is required. The compensation of axial clearance can be made very easily already as shown in fig 6.

FN[1] — From information supplied by Batelle Institute, Columbus, Ohio, USA.

The following design problems are important in vane machines:

## 1.   Bending loads of the Vanes

The vanes are subject to a circumferential force only if the pressure in both adjacent displacement chambers is different, if one of them is connected with intake and the other one with the delivery port. This force loads the vanes like a cantilevered beam in the slits of the rotor. The admissible bending stresses at the danger section at the beginning of the rotor slip determine the necessary thickness of the vanes. The details of these elementary calculations are not reproduced here, but it should be noted that Ernst, B3, recommends us not to exceed at this point a bending stress of 28 hebar (40,000 psi). The length of the vanes should be such that their supported length in the slot, in the outermost position, is at least 200% of the projecting length.

## 2.   Form of Track

The track of the vanes on the inner side of the stator ring must avoid sharp changes of direction leading to shock-like accelerations of the vanes. Preferably, the vanes should not move whilst they are loaded by the pressure between the intake and delivery distribution ports. Therefore, one gives the track a constant radius at these points, abandoning slightly the elliptical form.

## 3.   Contact between Vanes and Track

In order to avoid leakage, the tip of the vanes must always be in good contact with the track, and even with high speed pumps the centrifugal forces are sometimes insufficient. One method is to press the vanes outward with small springs applied at their rear, but such springs must make two strokes each revolution (in double vane pumps) and tend to be unreliable. Another method is to apply the delivery pressure on the rear side of the vanes by suitable depressions in the side walls. However, this leads to undesirably large contact forces and therefore some designs incorporate vanes that are divided either in the circumferential or in the longitudinal directions, with the pressure acting only on a part of the rear faces.

## 4.   Form of the Vane Tip

Normally one rounds the vane tip as far as possible, leading to a contact between a large radius and the track, thereby reducing the Hertzian stresses. Other designs use depressions on the tip of the vanes, filled with delivery pressure for transmitting the contact forces.

Apart from the usual form, with vanes in the rotor, other types of vane machines have radially moving vanes in the stator ring and an elliptical rotor, leading to similar design problems. Finally, for an excellent scientific treatment of vane machines, the reader is referred to the thesis of Wüsthoff /FN$^2$/

## 34–3    Other Hydrostatic Machines

There are many mechanisms which can be used in principle as displacement machines for incompressible lubricating fluids, (see, for instance B8). Most of them pose many difficulties to accurate manufacture, and only the screw pumps are industrially produced.

Fig 7 represents the longitudinal section through a screw pump with three rotors, where the two outer ones have milled-in grooves, similar to a steep worm gear. The central rotor driven by the input shaft carries a flange in the form of a helix, meshing exactly with the groove of the outer rotors. This produces displacement chambers

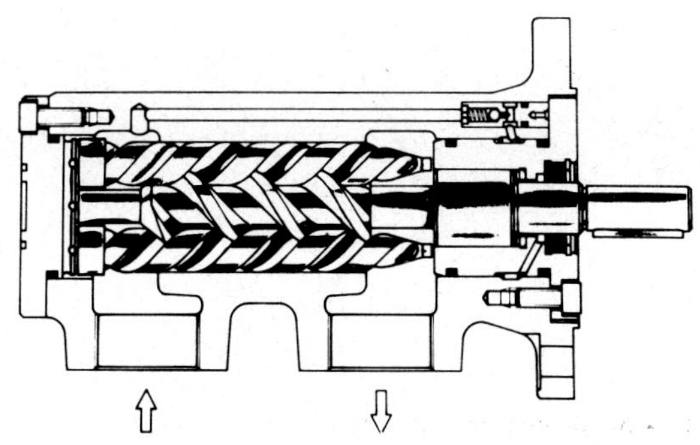

Fig 34.7: Cross-section through a screw pump (SIG)

FN$^2$ — Wüsthoff, Theorien und Messungen an hydrostatischen Flügelpumpen, thesis Eindhoven, 1969.

in the groove together with the contact of the faces of the rotor between themselves and the casing. On rotation of the rotors, the chambers travel axially, delivering fluid from an intake port at right to a delivery port at left.

Screw pumps have many critical dimensions including the profile of the grooves and of the flange, giving many difficulties on manufacture. They are, therefore, limited to relatively low pressures, up to about 70 bar (1000 psi), but they have the advantage of low noise operation.

They are manufactured only with constant displacement volume.

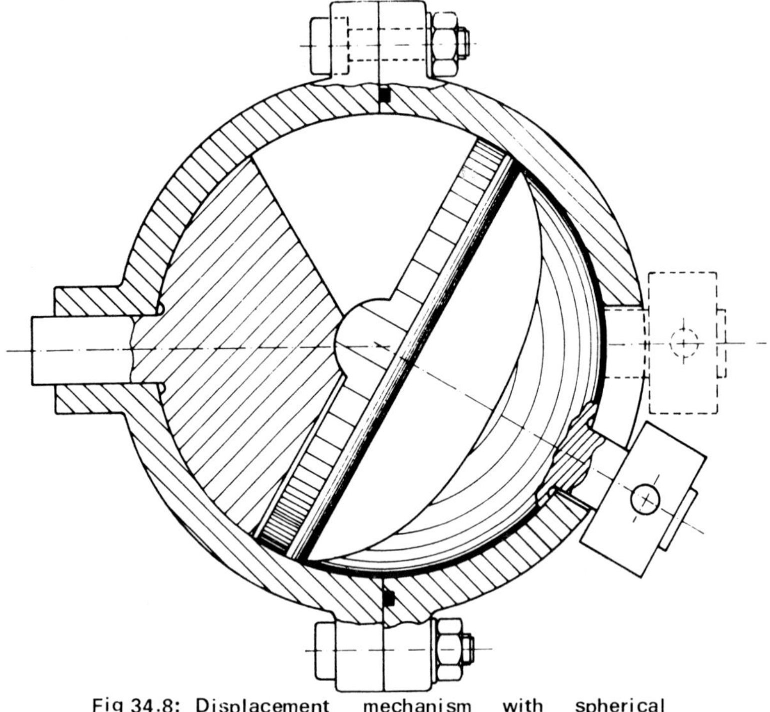

Fig 34.8: Displacement mechanism with spherical cavity

In conclusion, fig 8 represents an interesting but not industrially produced hydrostatic machine with a spherical cavity. It holds two shafts, similar to a universal joint, equipped with a disc between

FN[3] — They are also used to pump liquid or paste, like foodstuffs.

them touching the internal walls of the cavity for sealing. Depending on the tilting angle of the shaft at the right, the disc executes a dancing movement on rotation, periodically increasing and decreasing the displacement chambers between the disc and the internal spherical walls. Suitable distributing ports in the walls of the cavity, not shown, take care of the distribution.

Changing the tilting angle of the shaft (right) in the mechanism of fig 8 produces an adjustable displacement volume. This design has only surface sealing, but its main disadvantages are the large delivery fluctuations of the only two displacement chambers. The similarity to tilting head axial piston machines is also remarkable.

## 35    Model Laws and Operating Limits

### 35–1    Properties of Operating Limits

In Section 31 we have treated how rotation frequency and flow, torque and pressure depend on each other during the operation of hydrostatic machines. The characteristic variable was the displacement, adjustable in many piston machines. The inevitable losses constitute only small corrections to the fundamental equations 31–7 and 31–8, as we shall examine a little more closely in Section 36.

The operating limits, or maximum admissible values of the operating variables, have a quiet different nature. They are similar to legal prohibitions, as for instance the speed limits and the admissible loads in automobiles, imposed from the outside in the interest of safety. Similarly, as in road traffic, it is necessary to supervise the observation of the operating limits in hydrostatic machines, at least during the design phase of an installation. Hydrostatic machines can otherwise very easily attain dangerously high rotation frequencies and pressures, or even destroy themselves. Safety (pressure relief) valves are used frequently in hydrostatic installations for avoiding over-pressures, whilst corresponding limiters for the rotation frequency are used only in large machinery.

In particular, hydrostatic machines have the following operating limits:

1. The highest rotation frequency results from the admissible velocities of the different sliding surfaces.
2. The maximum pressure is determined by the admissible elastic stresses, and also by deformations that could lead to a dangerous reduction of certain gap heights or to high local edge or contact stresses with very small gap heights.

Starting from these limits, the maximum torques and flows are determined by the fundamental equations. We shall distinguish in Section 63 between the intrinsic operating limits, that cannot be exceeded by the machine in  practice, and between the external operating limits, any exceeding of which must be prevented by the installation.

The greatest admissible power, given simply as the product of displacement, maximum pressure and maximum rotation frequency according to equation 31–4, is sometimes asked for. We must not forget, however, that not the power itself, as in electrical machines due to heating of the copper, but rather the rotation frequency and the pressure are limited. If for any reason in an installation the maximum frequency cannot be used, then the maximum power becomes correspondingly smaller, since the pressure limit is unaffected.

Numerical example: An axial piston machine with $q_0 = 64$ cm³/rd ($q_0 = 402$ cm³/rev) has a maximum frequency of 160 rd/sec (1500 rpm) and a maximum pressure of 300 bar. This gives

$$\hat{Q}_{max} = q_0\, \omega_{max} = 64\ \frac{cm^3}{rd}\ \ 160\ \frac{rd}{sec} = 10\ 200\ \frac{cm^3}{sec}$$

$$\hat{M}_{max} = q_0\, P_{max} = 64\ \frac{cm^3}{rd}\ \ 300\ \frac{daN}{cm^2} = 192\ daNm$$

$$\hat{P}_{max} = 64\ \frac{cm^3}{rd}\ \ 300\ \frac{daN}{cm^2}\ 160\ \frac{rd}{sec} = 307\ kW$$

We must not forget that the maximum torque and flow can be obtained only with full displacement setting ($a = 1$), and that they must otherwise be reduced correspondingly. If we apply for instance on the machine of the above numerical example a torque of 192 daNm, it will produce with full displacement setting the admissible pressure of 300 bar (4 900 psi). Reducing the displacement setting say, to one-half, would increase the pressure, apart from several tens of bar for losses, to 600 bars (9 000 psi), probably rapidly destroying the machine. This is another example of the necessity to supervise the observation of the operating limits in hydrostatic installations.

The operating limits of a given design depend largely on the required reliability and life. Therefore, some manufacturers give different limits depending on the application. As an example, higher rotation frequencies and pressures are frequently admitted for vehicles, accepting thereby reduced life and reliability.

Some hydrostatic machines comprise parts with Hertzian contacts, where the relation between load, rotation frequency and life is similar to that in ball bearings. This leads to an increase of life by the factor eight with one-half operating pressure. Other parts have the normal behaviour of the material as in testing machines, but the real life until failure is always highly variable. Some manufacturers of long life axial piston machines claim that their machines do not have a natural life limitation, but 'die' only by accident. This can be true for hydrostatic bearings with sufficient gap height. Also lightly loaded roller bearings seem to have a life considerably longer than computed according to manufacturers catalogues, probably due to the excellent lubrication in hydrostatic systems.

The operating limits supplied by the manufacturers give therefore always only approximate indications. The real situation in an installation depends very much on incidentals or unpredictable changes of the load, like pressure peaks. On the other hand, one desires to increase pressure and rotation frequency as much as possible, in order to obtain as much power out of an installation as possible, thereby improving the economy.

Let us repeat for clarity that the operating limits have similar characteristics to the speed limits on the road. It is quite possible to exceed them a little, or even considerably, but the probability of getting caught increases rapidly. In the case of hydrostatic machines, this means usually almost total damage, since seizing of a sliding surface produces swarf and steel particles that will be carried by the oil to other sliding surfaces and probably destroy them.

In practice, completely destroyed hydrostatic machines are sometimes presented, where the internal parts are milled down into small particles and almost unrecognisable. This happens if at the beginning of trouble the machine is not quickly shut down, because even in normal operation the pressure and stresses are relatively high. Sometimes one needs almost a criminal eye, in order to find the guilty party, that is the part where the trouble started.

Apart from the limits for rotation frequency and pressure, there are other regions of operation that must be avoided. Some hydrostatic machines with adjustable displacement have a tendency to overheat if they are run with high rotation frequency but small displacement setting, especially with high operating pressures. This corresponds to a very small flow, as needed if in a hydrostatic winch drive the load is suspended without movement.

Apart from these limits due to safety, there are other regions where operation is not advisable because of high relative losses. These include, especially with axial piston machines, operating pressures of less than 50 bar (700 psi), since then the loss pressure, as defined in equation 31–6, is relatively high.

Some manufacturers permit higher pressures only for a part of the operating cycle, indicating for each pressure the admissible operating time as fraction of the entire working time. As an example in an axial piston machine the pressure of 150 bar can be supported continuously, and of 250 bar (3 800 psi) only during 25% of the operating cycle. This method is satisfactory in practice, in order to obtain an industrial life of 6 000 to 12 000 hours, corresponding to three to six years in single shift operation. But such hydrostatic machines can run with higher limits, accepting correspondingly reduced life.

## 35–2   Model Laws

Hydrostatic machines, especially axial piston machines, are frequently built in a range of models,where all the dimensions in each model are simply increased by a certain factor. We shall, therefore, examine here the situation of such a range.

Due to the high operating pressure in hydrostatic machines, the pressure forces exceed by far all other forces, and determine the range of models. We shall see that the stresses remain equal in all component parts, if the models of the range are geometrically similar.

We consider now a range of hydrostatic machines, obtained from a basic model by multiplying all lengths with a factor $\lambda$. This leads to the following conditions.

1.  All angles remain without change.
2.  All lengths increase with $\lambda$.
3.  All areas are proportional to $\lambda^2$.
4.  The volumes, and especially the displacement volume, is proportional to $\lambda^3$.
5.  With the same mass density, or if the corresponding parts are made from the same material, the mass and the weight increase proportionally to $\lambda^3$.

If we run our model range with the same pressure, then we have furthermore:

6.  All pressure forces including the piston forces are proportional to $\lambda^2$.
7.  All moments or torques are proportional to $\lambda^3$.
8.  The elastic stresses are independent of $\lambda$.

The above relations are true for tensile and compressive stress, because both forces and cross-section are proportional to $\lambda^2$, and also for bending stresses, since bending moments and resisting moments are proportional to $\lambda^3$. This is strictly true as long as the weight can be neglected compared to the pressure force, which is normally the case. On the other hand, the weight increases with $\lambda^3$ more quickly than the pressure forces increasing with $\lambda^2$, and it can become considerable with very large machines.

9. The elastic deformations are proportional to $\lambda$.

Furthermore, if we let run our model range with the same rotation frequency:

10. The velocities, including the piston velocity and the flow velocities are proportional to $\lambda$.

11. The flows and the transmitted power are proportional to $\lambda^3$.

12. The power density, that is the ratio of power and mass is independent of $\lambda$.

13. The acceleration and centrifugal forces are proportional to the mass, to the square of the rotation frequency, and to a radius, therefore proportional to $\lambda^4$ (or $\omega^2\lambda^4$).

Since the elastic stresses are independent of $\lambda$ it is possible to admit the same maximum pressure for the entire model range. On the other hand, in order to maintain the sliding velocities constant for the same security against seizing, it is usual to reduce the frequency limit of our range proportionally to $1/\lambda$. Under these conditions the flow and the maximum power increase only with $\lambda^2$, and the power density decreases with $1/\lambda$, or larger machines need more mass compared to their power. The acceleration and centrifugal forces become therefore proportional to $\lambda^2$ (because $\omega^2 \lambda^4 \sim \lambda^2$) and their ratio to the pressure forces remains constant.

The conditions of variation must be indicated very precisely, including the behaviour of the operating limits, for the determination of the model range. This depends on the question which variable or physical law determines the security of operation. In very large machines, in hydrostatics with a displacement volume of more than 5 lit/rev the weight becomes important and imposes a reduction of the maximum pressure.

*Model laws for the losses*

The losses depend essentially on the gap height and have therefore a complicated behaviour. As a practical indication we can say the following:

*1. Torque losses*

The torque losses increase proportionally to $\lambda^3$, that is proportionally to the displacement volume, and therefore the loss pressure introduced in equation 31–6 remains independent of $\lambda$. With the same operating pressure, the relative power losses and the efficiency are constant.

## 2. *Leakage losses*

According to the gap formula the leakage losses are proportional to $h^3/FN^1/$. Choosing a gap height proportional to $\lambda$ gives a leakage flow corresponding to $\lambda^3$, but with reducing rotation frequency ($\omega \sim 1/\lambda$), the relative loss and the volumetric efficiency decreases. The selection of the gap height is governed by our remarks of 25–6, but in practice, the gap height increases usually a little less than with $\lambda$, so that the volumetric efficiency remains constant.

### *Specific rotation frequency*

Since the admissible sliding velocities are limited by the different surfaces, the possible rotation frequency depends on the absolute size of the hydrostatic machine. In order to be able to compare the maximum rotation frequency of hydrostatic machines with different displacements, we introduce a specific rotation frequency. All sliding velocities are proportional to the specific rotation frequency, independently of the maximum displacement of the machine. It is therefore proportional to the third root of the displacement volume (this being proportional to the length factor $\lambda$) and to the rotation frequency. For convenience one associates a specific rotation frequency of *one* to the usual values of the displacement volume of 16 cm$^3$/rad and to the rotation frequency 160 rad/sec (or 100 cm$^3$/rev and 1500 rpm), obtaining the formula:

$$\Omega = \frac{\omega}{160 \text{ rd/sec}} \sqrt[3]{\frac{q_0}{16 \text{ cm}^3/\text{rd}}}$$

(35–1)

For general classification we have

| | |
|---|---|
| $\Omega < 1$ | low rotation frequency |
| $\Omega = 1 \cdots 1.5$ | medium rotation frequency |
| $\Omega > 1.5$ | high rotation frequency |

As an application of the specific rotation frequency, the low speed oil motors treated in Section 32–4 have normally a specific

---

FN$^1$ — The leakage flow is independent of $\lambda$ with constant gap height.

rotation frequency of 0.1 up to 0.4 with a displacement of $q$ = 100 to 1200 cm³/rd.

As a conclusion from our model laws, a range of hydrostatic machines is preferably built by geometric similarity from a base model, as long as pressure forces are preponderant. The elastic stresses are then the same in all sizes, and if the machine is running with the same specific rotation frequency, also the sliding velocities. In practice, the geometric similarity is slightly disturbed by the use of normalised parts, as, for instance, screws. It is then convenient to set up the range of models the sizes as a geometric progression, meaning that each model is obtained from the preceding one by multiplication with the same factor, called the size factor.

The selection of a suitable size factor involves a compromise between the following:

1. With a small size factor, there is a model more closely suited for each application.
2. With a large size factor less models, tools and spares are needed in order to cover a given range.

Balancing these conflicting requirements, a size factor of the displacement volume of 2 to 2.5 seems to be appropriate for oil hydraulics. On the other hand, special intermediate models will always be considered for applications with a large number of machines.

So far we have treated the model theory more from the practical angle. In principle, one should set up dimensionless parameters or product groups of the variables and study how they change over the model range. There are several product groups (see B 14, Section 16), but in practice the following are the most important:

$$\frac{\sigma}{p} \qquad \frac{p}{E} \qquad \frac{p}{B} \qquad \frac{\mu\omega}{p} \qquad\qquad (35-2)$$

Equation 35–2 invites the following comments:

1. The first expression gives the elastic stress as a multiple of the operating pressure. With axial piston machines it is limited to about 5 to 6 for tensile and compressive stresses, giving conditions for the dimensioning, as, for instance, the ratio of the diameter of the connecting rods and the pistons.

2. The second expression is the ratio of the operating pressure
   to the modulus of elasticity (Youngs modulus) and therefore a
   measure for the elastic deformations.

3. The third expression is the ratio of the operating pressure to
   the compressibility modulus, and determines the importance
   of the compressibility effects, as, for instance, the admittance
   of the metering compression chambers (see Section 33–2).

4. The last product group compares the operating pressure with
   the viscosity forces. It is needed for the friction losses of
   hydrostatic bearings (equation 25–3) in hydrodynamical bearings, by setting $v = \omega r$ in formula 26–1, and also in mathematical models for the efficiency.


The above parameters are multiplied in many cases, by the ratio.
of two lengths, in order to obtain the real values, for instance of the
deformation, as used in Section 72. In this connecting the investigations about the model theory of hydrostatic machines of A. Teslenko,
Moscow, should be mentioned /FN[2]/ Teslenko considers the question
of an optimal extrapolation of given experiences for setting up a
range of hydrostatic machines based on model theory.

FN[2] – Different Russian publications and an article: 'Théorie de la similitude pour
la construction des machines volumetriques' in 'Energie Fluide', Paris 1969.

## 36    Characteristics of Hydrostatic Machines


After treating the operating limits in Section 35, we shall examine here the actual values of the operating variables in hydraulic installations. The fundamental relations for hydrostatic machines are given by equation 31—7 and 31—8, but the actual values are determined by interaction with the installation.

Fig 1 shows a very simple circuit with an adjustable displacement hydrostatic machine at the left, able to deliver hydraulic fluid in both directions to the throttling device at the right. In order to avoid cavitation both conduits between the hydrostatic machine and the throttle are connected over non-return valves to feeding lines from the oil tank.

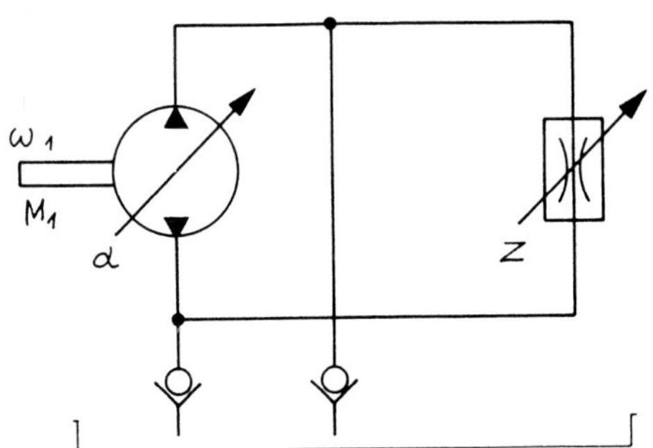


Fig 36.1: Closed   circuit   for   the   determination   of
characteristics

Beginning an experiment with completely opened throttle (Z = o), and with a fixed value of rotation frequency and displacement setting, the flow will run through the throttle without pressure drop. The hydrostatic machine needs now only its loss moment, in addition to a small torque corresponding to the pressure drop in the conduits including any remaining pressure drop of the opened throttle. Begining to close the throttle, its impedance, the pressure drop and therefore the torque drawn by the machine will increase. With sufficiently high impedance, dangerous overpressures may develop. It is therefore advisable for safety to put a relief valve in parallel to the throttle.

The functional diagram of fig 2 explains these relations more in detail. The input variable on the block above to the left is the rotation frequency $\omega_1$ /FN[1]/ producing the ideal loss free flow $Q_i$ according to equation 31–1. This equation is entered in this block, and the influence of the displacement setting represented by the lateral connection.

Deducting from the ideal flow the leakage flow $Q_S$ we obtain the effective flow entering the throttle. It produces the pressure p, as represented by the block at right with the impedance Z, depending on the throttle setting. The operating pressure is responsible for the leakage flow over the central block with Y and for ideal torque M; by the lower block, according to equation 31–2. Adding the loss torque, one obtains the effective torque that the hydrostatic machines opposes to its rotation.

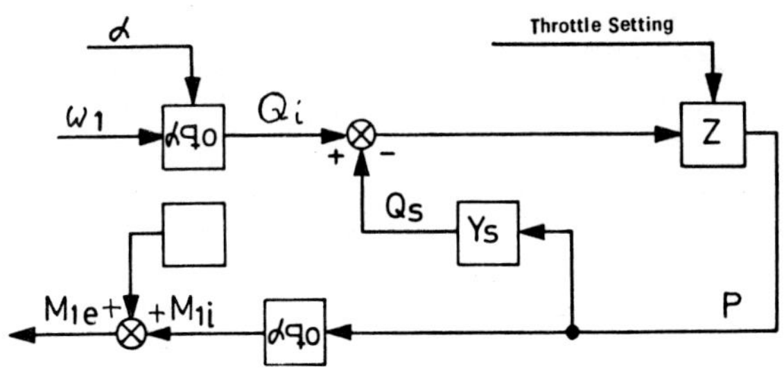


Fig 36.2: Functional  diagram  of  the  circuit  of  Fig.1

We note the following important points on fig 2:

1.  The development or the actual values of flow and pressure depend not only on the hydrostatic machine but also on the impedance of the throttle. Particularly, closing the valve can produce dangerously high over-pressures, unless the correspondingly high input torque stops the drive of the hydrostatic machine.

2.  The relation between torque and rotation frequency of the hydrostatic machine, important for the selection of its drive, depends on the impedance of the throttle (together with the displacement setting and the maximum displacement).

FN[1] – We choose the symbol $\omega_1$, since we shall use in hydrostatic machines $\omega_2$ for the output frequency in Section 62.

Whilst overpressures can be avoided by suitable control of the impedance or by a relief valve, it is also necessary to avoid excessive rotation frequencies. These could be produced with an imperfectly governed drive and completely opened throttle, because the hydrostatic machine then needs almost no torque. This is an interesting illustration of the fact mentioned in Section 35, that the operating limits are distinct from the operating variables and must be supervised carefully.

The functional diagram in fig 2 is valid also with negative displacement se.'ing and rotation frequency, producing a negative flow and pressure. A negative flow means simply a flow in the inverse direction, whil. a negative pressure means that the absolute pressure in the lower m.in conduit of fig 1 is higher than in the upper one. The non-return valves and the tank (with a free surface) prevent negative absolute pressures and cavitation in both main conduits.

After describing our simple installation by a circuit (fig 1) and by a functional diagram (fig 2), we shall now examine a third way of representing its operation, the field of characteristic curves or characteristics in fig 3. Here we use, in analogy with the servovalves in Section 45, the Y representation, with the pressure on the x axis and the flow on the y axis. With constant rotation frequency, we obtain the slightly inclined characteristics between pressure and flow, one for each displacement setting. The inclination corresponds to the leakage admittance which was represented by the central block in fig 2. .

Fig 3 contains furthermore the secant admittance of the delivered flow, corresponding to the inclination of the secant between the origin and the operating point. It is seen how it depends on the operating point, and especially on the displacement setting. Normally the inclination of the secant is much larger than the inclination of the characteristics due to leakage, but it may become comparable with very small displacement setting.

Fig 3 contains also characteristics with negative pressure and flow, as obtained by machines with two directions of flow by negative displacement setting /FN²/. Reversing the direction of flow reverses also the direction of pressure drop on the throttle, so that the power always flows from the machine to the throttle.

---

FN² — or with positive displacement setting, by negative rotation frequency.

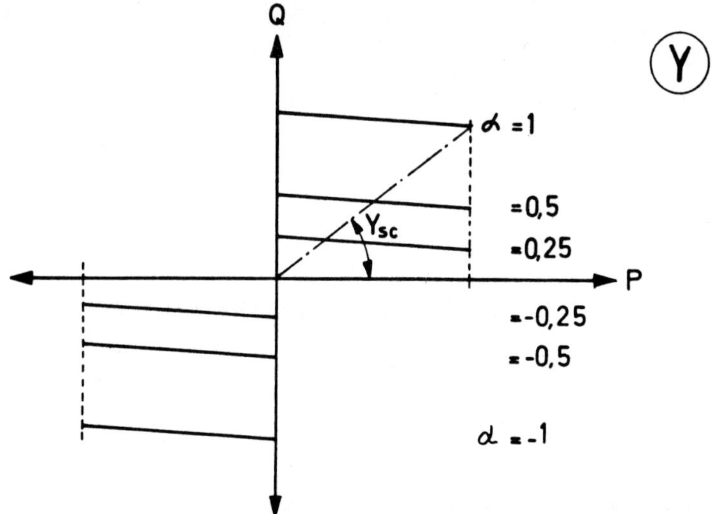



**Fig 36.3:** Characteristics of a hydrostatic machine with
adjustable displacement

We could now ask the question, whether it is possible to extend
the characteristics in fig 3 also to the left above and to the right
below, corresponding to positive flow and negative pressure, and to
negative flow and positive pressure. This is quite possible in principle, only then power is delivered by the load into the hydrostatic
machine working as a motor. Instead of a throttle, we would have to
use another consumer, for instance, a hydrostatic machine driven
exceptionally by its output.

This is the case when the hydrostatic transmission brakes the
load through the output shaft.

Fig 3 also contains the flow and pressure limits, where the
maximum flow is determined by the full displacement setting. It
cannot be exceeded with fixed rotation frequency, representing
thereby an intrinsic limit in the sense of Section 35, whilst observation of the pressure limits must be assured by proper control of
the throttle. It is therefore an external limit, and we shall use it in
Section 63 for the dimensioning of hydrostatic transmissions.

Similar characteristics can be produced by hydrostatic machines
with fixed displacement, by changing the rotation frequency and
perhaps also the direction of rotation. The actual operating pressure
is then again determined by the impedance of the throttle.

# 4   HYDROSTATIC COMPONENTS

Whilst hydrostatic machines convert mechanical and hydrostatic powers, other components are needed for controlling, direction and processing the flow. They comprise the different valves and also some auxiliary parts which we shall describe here. The valves can be classified by the number of connections as follows:

   a) Dipole valves
   b) Valves with several connections

On the other hand, valves can be classified by their function as follows:

   1) Throttle valves
   2) Relief valves
   3) Flow control valves
   4) Directional control valves
   5) Servo valves

## 41   Throttle Valves

### 41–1   Introduction

Throttle valves are components installed in hydraulic conduits for producing a pressure drop. Basically, they are devices with narrow cross section, losing pressure by high flow velocity. Normally, the cross section and, thereby, the impedance and admittance of the throttle valve can be adjusted.

According to their form, throttle and other valves can be classified as follows:

1) Seating valves
2) Sliding valves

Both forms produce the adjustable throttling section between a movable part, the valve body, and a casing, as shown schematically in fig 1. In seating valves, at the left in fig 1, the valve body rests on a seat in the casing when closed. In the sliding valve, at the right, the valve body closes the cross section by passing over a port or a circular groove without coming to a definite stop. The edges limiting the narrow cross sections are called the metering edges.

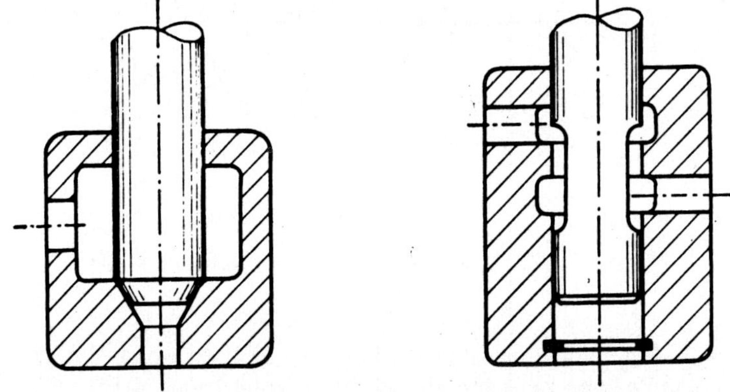

**Fig 41.1: Principle of seating and sliding valves**

## 41-2  Forces on Seating and Sliding Valves

Seating valves have less leakage flow in the closed position (or their admittance is very small) and they are not so sensitive against contamination in the oil, but they need much higher forces. The actuation force of the valve body, in fig 1 at the left, is approximately equal to the product of the pressure and of the mean cross section of the conical seat. From this, the force generated by the stem of the valve when leaving the pressure chamber, in fig 1 above, given by the product of its cross section and of the pressure must be deducted.

In principle,the forces can be compensated by a suitable selection of the cross sections of the valve seat and of the stem but, in practice, a highly variable net force remains.

Sliding valves  in the form represented in fig 1, are free from hydrostatic forces (forces due to the static pressure difference). On the other hand, high flow velocities exist between the metering edges, transmitting momentum and producing dynamical forces on the valve body.

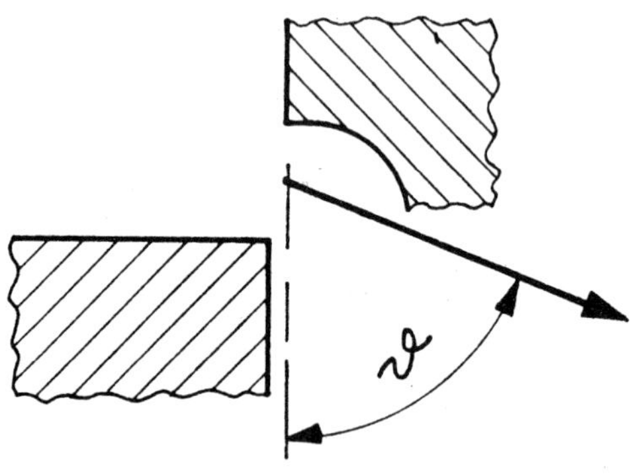

Fig 41.2: Metering edges of a sliding valve with jet angle

Fig 2 represents the metering edges on an enlarged scale. The oil jet leaves under an angle $\theta$ with respect to the axis of the valve body, producing a total force in the direction of the jet, given by formula 13–9. The radial components compensate each other in round and other cross sections of valve bodies and with symmetric arrangement of the ports in the circumferential direction, whilst the axial components are additive, leading to

$$F_a \;=\; 2\,A\;\frac{\Delta p}{\zeta}\;\cos\theta \qquad\qquad\qquad (41\text{–}1)$$

with A = open cross section

$\Delta p$ = pressure drop

$\zeta$ = loss coefficient ($\sim$2)

$\theta$ = angle between jet and axis of the valve

The angle of the jet $\theta$ depends on the relation between the radial clearance of the valve body in the casing and the height, ie the extension along the axis of the opening. We have :

$\theta = 69°$ with large opening height

$\theta = 45°$ with opening height equal to radius clearance
        (follows from symmetry)

$\theta = 21°$ with small opening height

With a large opening, we therefore obtain the following, relatively high force, with $\cos 69° = 0.36$ and $\zeta = 2$,

$$F_a{}' = 0.36 \, A \, \Delta p \qquad\qquad (41–2)$$

It should be noted that the magnitude and the direction of the force on the valve body remain the same if the direction of the flow is inverted, because the jet then transports the inverse momentum out from the space of the valve. The direction of the force tends therefore always to close the valve.

With inverse flow, the pressure drop in formula 1 becomes negative, although the force maintains the same direction. In principle one should, therefore, write $\underline{\Delta p}$ absolute in formula 1 (omitted here for simplicity). If the force is expressed by the square of the mean flow velocity, corresponding to formula 13–8, this fomal difficulty would disappear.

There are many methods of reducing the momentum forces on the valve body like giving a special form to the part where the jet impinges, as described by Guillon, B 7, Section 4–3.

## 41–3   Characteristics of Throttle Valves

In practice, many throttles have the form of needle valves, as represented in fig 3, allowing a very fine adjustment. For high impedances, a valve body with a very sharp cone in the form of a needle is needed.

The pressure/flow characteristics depend on two causes for the pressure drop :

1.  The viscosity forces
2.  The inertia (turbulence) forces

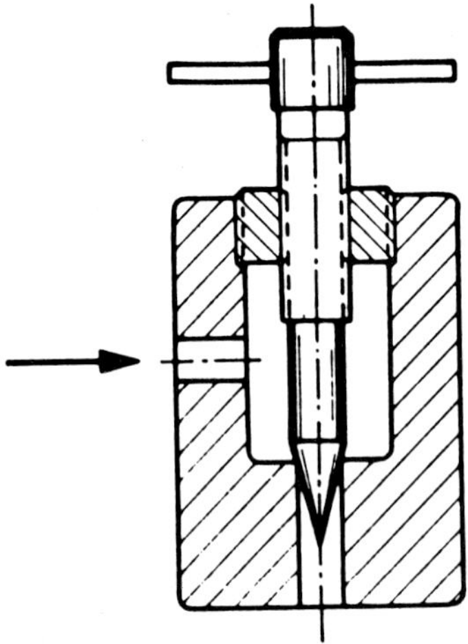

**Fig 41.3:** Schematic cross-section of a needle valve

The viscous forces give a linear relation between flow and pressure drop as we know from the gap formula. Such throttles have long and narrow channels, but also the disadvantage of reducing their impedance greatly with increasing temperature, due to the decrease of viscosity.

The inertia forces are independent of the temperature and give a pressure drop proportional to the square of the mean velocity and the flow. Throttles with a very short, narrow cross section, such as diaphragms, have such a charateristic with preponderant inertia forces.

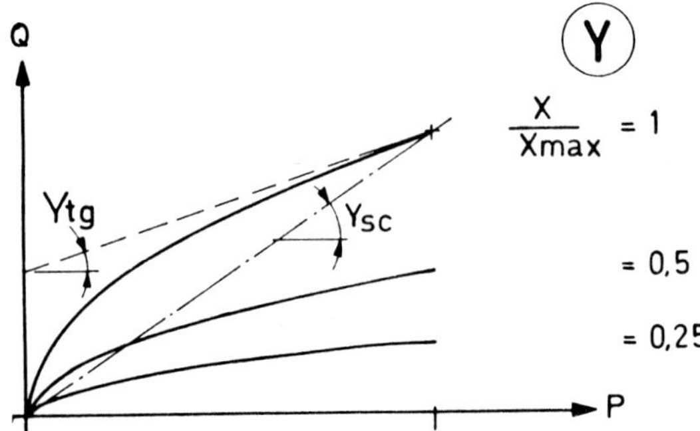


**Fig 41.4: Characteristics of a turbulent throttle**

Fig 4 represents the characteristics of an adjustable throttle with preponderant inertia forces in Y representation for three different positions of the valve body, namely with full, one half and one quarter opening. The tangent and secant admittances, in the Y representation proportional to the inclination of the tangent and the secant, are entered in fig 4 for an operating point with full opening and full pressure drop. It is seen in fig 4 that the secant admittance is twice as large as the tangent admittance. /FN[1]/ This is true for all operating points on a turbulent characteristic which follows from the properties of a parabola. On the other hand, the admittances depend both on the opening and the applied pressure.

An adjustable throttle is thus a dipole, as discussed in Section 14. Pressure or flow can be freely chosen but then flow or pressure are determined with given adjustment or setting of the valve body. /FN[2]/.

FN[1] — Conversely, the secant impedance is half the tangent impedance.


FN[2] — In a certain sense, we have here three variables of which two, namely adjustment and pressure or flow, are free and the other one determined. In other words, an adjustable throttle is a component with two degrees of freedom.

## 42   Pressure  Relief  Valves

Pressure relief valves are used frequently to maintain a pressure constant or below a selected maximum value. Whilst there are many different forms, they consist essentially of a spring-loaded valve body closing an opening. When the pressure in the opening reaches the selected or set value, as given by the spring force and the effective cross section of the valve body, it lifts and opens a passage for the fluid. Non-return valves are a special kind of pressure relief valve \ opening on a very small pressure but blocking the flow in the inverse direction. They have a similar layout, but the spring is made very weak or missing altogether.

Pressure relief valves can be classified as follows:
1.  Direct acting valves
2.  Non-return valves
3.  Piloted valves
4.  Special designs

### 42–1   Direct Acting Pressure Relief Valves

Direct acting pressure relief valves are built both as seating and as sliding valves and the valve body can take many different forms. The most simple seating valve comprises a sphere pressed onto a seat by a spring. The static and dynamic forces are the same as discussed in Section 41.

Fig 1 represents a pressure relief valve with a valve body having a conical part collaborating with a seat in the casing. It is loaded from above by a spring and the opening pressure of the valve can be set or selected by changing  the preload  force of the spring with the handwheel and the thread above. The fluid enters laterally in the lower part, flows between the valve cone and the seat and leaves the casing through the lateral bores in the centre.

Notable in fig 1 is the extension of the valve body below terminating in a small piston in a cylindrical bore. This produces friction forces on the valve body damping any tendency to instability or vibration. At the same time, it has the advantage that a large part of

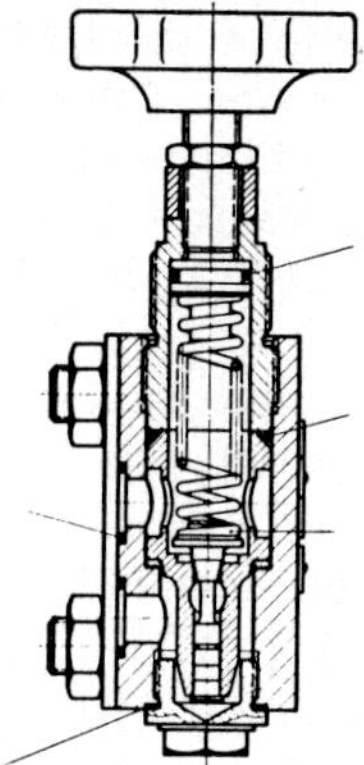

Fig. 42.1: Cross-section through a direct acting pressure
relief valve (Herion)

the pressure forces act on the lower surface of the piston where it is
not influenced by the high flow velocity through the valve seat.

We use the functional diagram in fig 2 in order to study the operation of such a valve more in detail using the Y representation.

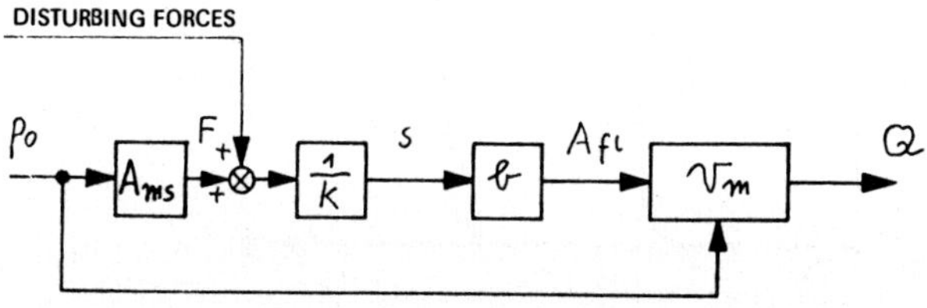


Fig. 42.2: Functional    diagram    for    the    direct    acting
pressure relief valve

The applied pressure p is the input variable in fig 2 determining
the force on the valve body by multiplication with its effective area.
/FN[1]/ This force determines, together with the disturbances like
friction or the momentum forces treated in Section 41, the position s
of the valve body. This is indicated by the next block with the
symbol l/k, the reciprocal spring constant.

FN[1] — This area is equal to the cross-section at the position where the cone rests on
the seat.

This block gives a linear relation between input and output, but with a threshold corresponding to the characteristics of preloaded springs. The displacement of the valve body is zero up to the pre-loading force and increases proportionally to the pressure increase thereafter. This is the case with coil springs, while the so called dished springs have a non-linear characteristic.

The displacement of the valve body determines, with the cross section $A_{fl}$, the opening by the next block with the opening width b and, furthermore, the flow as indicated by the last block. This is represented by the mean flow velocity $v_m$ depending on the pressure drop as discussed in Section 13. With the usual pressures in oil hydraulics, we obtain high flow velocities of the order of 50 m/sec and therefore, the turbulent pressure drop is preponderant.

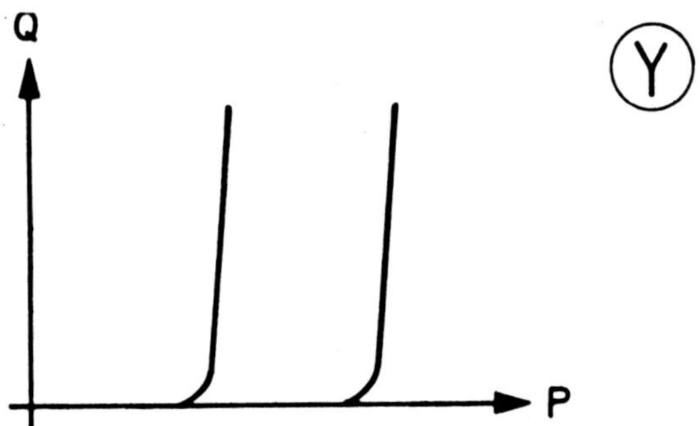


Fig. 42-3: Characteristics of an adjustable pressure
relief valve in Y representation

Fig 3 represents the characteristics in the Y representation for two different settings. The opening pressure depends on the preload of the spring. Furthermore, a softer spring produces steeper characteristics or higher tangent admittances as can be seen from the functional diagram in fig 2, since then the value of the spring constant k becomes smaller. The reduction of the incremental spring forces has a disadvantage in that the disturbing forces, especially the momentum forces of oil jets,gain more relative importance, leading to instability. Any artificial damping forces will, therefore, be useful, like the friction forces of the small piston in fig 1.

It is seen from the functional diagram that the sensitivity or the tangent admittance of the characteristics can be increased by increasing the effective area $A_{ms}$ of the valve body. However, this leads to large forces and to a bulky and expensive spring. /FN$^2$/

An ideal pressure relief valve would have vertical characteristics in the Y representation or an infinite tangent admittance corresponding to a pressure drop independent of the flow. This is not possible due to the instabilities and to the spring, but piloted relief valves bring an important improvement in this direction, especially for large flows.

Another important characteristic of the relief valve is the time needed to open it under a sudden pressure increase such as a pressure shock wave. This response time is usually of the order to 100 msec. In order to shorten the response time, the valve can be made sensitive not only to the pressure but also to the change or time derivative of the pressure, but these interesting questions are outside the scope of this book.

### 42–2    Non-Return Valves

Non-return valves should pass the flow in the opening direction with little or no pressure drop and completely block it in the other direction. Basically, one uses the same design as for relief valves but without the spring. Sometimes, a very weak spring is used in order to prevent the valve body from leaving the seat completely.

A non-return valve is, therefore, a relief valve set to zero pressure and the layout of fig 1, less the spring, can be used very well for this purpose. In the blocked direction, high pressures will be taken up with a minimum of leakage flow.

Disturbing forces have an especially large influence on the position of the valve body since the spring forces are missing. Therefore, non-return valves are more subject to instabilities. With spherical valve bodies, instabilities give rise to high frequency vibrations or noise and artificial damping is normally needed.

---

FN$^2$ — This is an example of the help given by functional diagrams for examining design changes.

The so called pilot controlled (or piloted) non-return valves are an interesting variant. Applying a control pressure to the pilot conn- ection lifts the valve body into the open position, allowing a flow in both directions. Without pilot pressure, it acts as a normal non-return valve.

### 42-3   Piloted Valves

Piloted valves are especially suitable for high flows or high tangent admittances. Basically, they are a parallel combination of a small pressure relief valve and a throttle valve, usually in the form of a sliding valve, controlled by the small relief valve.

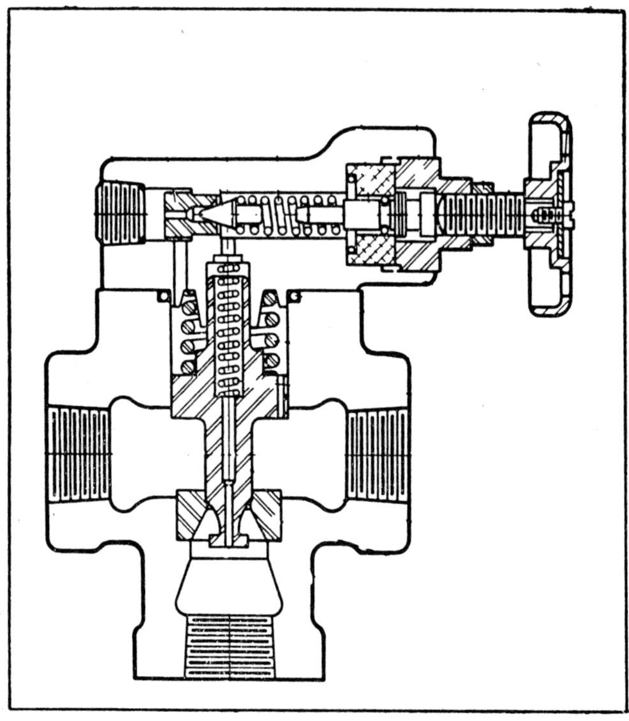

Fig 42.4: Cross-section through a piloted pressure relief valve (Vickers)

Fig 4 represents a piloted pressure relief valve with the small valve, also called a pilot valve, arranged horizontally on top, similar to fig 1. The main valve has a lateral entrance port and a vertical main valve body, which can move upwards, opening a passage below. The main valve body carries a differential piston in a cylinder with a spring. The external cylinder space is filled over a narrow duct in the differential piston with throttled pressure oil from the entrance connection and connected by another duct with the pilot valve. Initially, input pressure is in the cylinder, pressing the main valve body below and closing the passage below. If the pilot valve opens after reaching its set pressure, the pressure in the cylinder decreases, since the oil must be replaced by the flow through the throttled duct. The main valve body lifts and opens the downward passage below.

A detail of fig 4 is that the flow of the pilot valve is conducted through the axis of the main valve into the exhaust. Furthermore, it is possible to apply another, perhaps remote or distant, pilot valve that can be set by hand or electrically, giving a remote control of the entire pressure relief valve. Finally, the small deflector plate at the lower end of the vertical valve body compensating the forces of the outgoing jet, should be noted.

The functional diagram in fig 5 is a more precise representation of the pilot operation. From the applied pressure $p_0$ as input, we obtain the pressure in the pilot valve by subtraction of the pressure drop of the throttle, as described by the first block with the admittance $Y_{v1}$. The output is the small flow $Q_1$, determining through the impedance of the throttle, the pressure drop $\Delta p$ that reacts on the input. This pressure drop gives, together with the area of the valve body $A_{ms}$, the force that determines (with the disturbing forces) first the position of the main valve body and then the area $A_{fl}$ of the opening, as described by the common block with $b/k_2$. The last block determines the turbulent flow in the output.

The functional diagram again shows the influence of the design parameters. For a high gain, it is possible to select a soft spring for the main valve body ($k_2$ small) but then the disturbing forces increase their relative importance.

Piloted pressure relief valves are very useful in practice due to their steep characteristics or large tangent admittances. Such valves are also often called pressure control valves, a term that can be explained best with the Z representation of the characteristics. In this representation, the valve transforms any variations of flow into much smaller variations of pressure. In this sense, it produces an approximately constant or, at least, a much less fluctuating pressure. According to information supplied by the manufacturer of the valve in fig 5,the ratio of tangent to secant admittances(at maximum pressure) goes up to the value of 25 with small (nominal opening 10 mm) and to the value of 100 with large (nominal opening 32 mm) models.

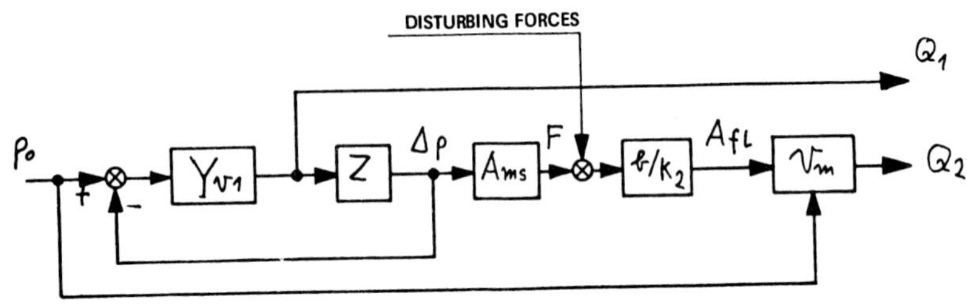


Fig. 42-5: Functional   diagram   of   a   piloted   pressure
relief   valve

## 42–4   Special Designs

Special forms of pressure relief valves are built with different connections of the piloting valve.It can be installed not only between entrance and outflow but also in other conduits. In practice, such valves have very similar design details, for which the reader is referred to B 2.

## 43    Flow Control Valves

Flow control valves are in essence the opposite of pressure relief valves, since they produce large variations in pressure drop with small variations of flow. The tangent admittance of the operating point is, therefore, small compared to the secant admittance.

The operation of flow control valves is best studied in Z representation with given flow and with the pressure drop as a result of the valve action. Flow control valves have essentially a series connection of a (frequently) adjustable throttle and a main sliding valve, normally kept open by a spring, resulting in the large tangent impedance. The pressure drop on the throttle valve is conducted to the end faces of the main slide valve and closes after a certain set or threshold value, producing thereby an additional impedance for the flow. /FN[1]/

Fig 1 represents a cross-section through an industrial flow control valve. The throttling valve at the left has the form of a sliding valve with adjustment by a knob and a thread. In order to reduce the gain between position and impedance of the throttle valve, the valve body

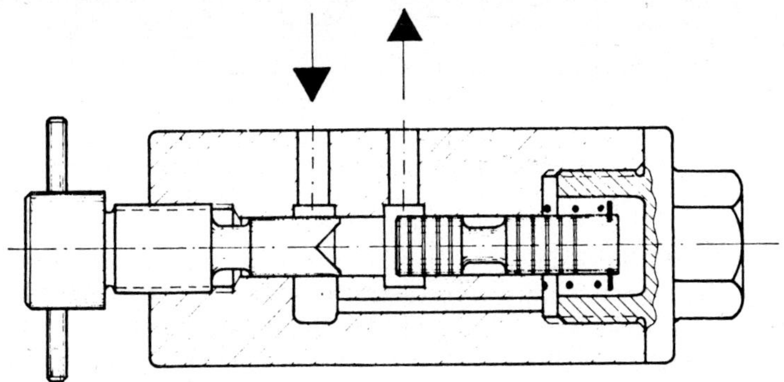

Fig 43.1: Schematic cross-section through a flow control valve

---

FN[1] — The main slide valve is similar to the piloted relief valve, only it is connected in series and not in parallel to the throttle valve.

is equipped with two triangular grooves through which the oil flows when in a half-closed position. Afterwards, it passes the first edge of the main slide valve, producing a further pressure drop and leaves the casing. The main valve is pressed by the spring to the right into the open position, but the duct below carries the pressure from upstream of the throttle valve at the rear-end face of the main valve. Since the front-end face is directly downstream of the throttle valve, a net force corresponding to the pressure drop presses the main valve to the left. It closes the flow cross-section depending on the spring characteristics, and produces the control impedance.

For a more detailed study, we use the functional diagram in fig 2, with the flow as input variable, producing the pressure drop of the throttle valve by the first block with the impedance $Z_{v_1}$. It acts on the end faces with the area $A_{ms}$ of the main valve, producing a force to which the disturbing forces are added. The third block produces the position $s_2$ of the main valve with the spring constant $k_2$ for small movements. Instead a preload of the spring produces a threshold value so that the main valve begins its travel only after a certain force is reached.

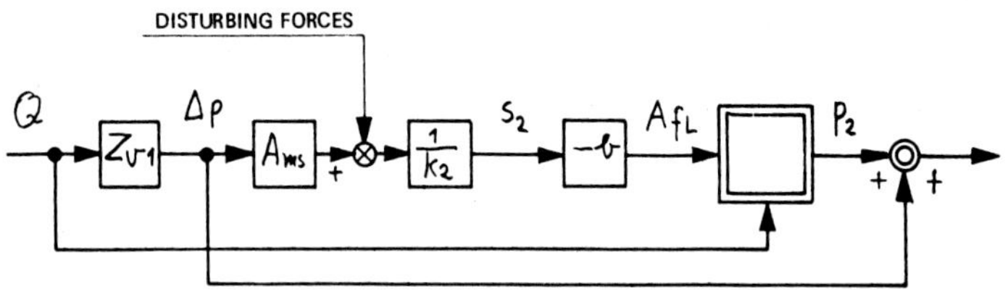



Fig 43.2: Functional diagram of a flow control valve

The fourth block in fig 2 gives the relation between the position of the main valve and the cross-section of the opening denoted by -b, since a positive movement reduces the cross-section. The cross-section determines the pressure drop of the main valve, depending on the flow as indicated by the last block. Adding finally the pressure drop of the throttle valve, we determine the pressure of the entire flow control valve. The last block has been drawn with double lines since it indicates the essentially non-linear relation between opening and pressure drop. It is governed by inertia forces and very well represented by the former formula 13–5 referred to the actual opening cross-section with a loss factor of $\rho = 2$. Some of the other blocks are also more or less non-linear, as can be described by the distinction of tangent and secant impedances.

Fig 3 represents a practical flow control valve similar to fig 1. The external knob allows us to set the opening pressure of the throttle valve and thus the flow where the main valve begins to close.

**Fig 43.3: Adjustable   flow   control   valve   (Wandfluh)**

Fig 4 represents the characteristics of a flow control valve. /FN$^2$/ The lower part of the characteristics is governed by the throttle valve with approximately quadratic pressure increase, until the

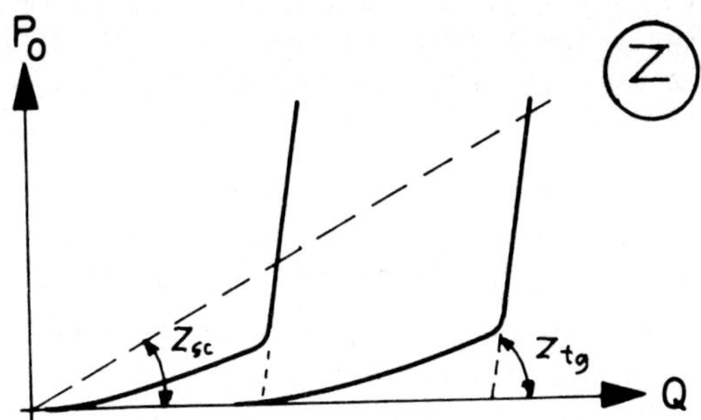


**Fig 43.4: Characteristics of a flow control valve**

F N$^2$ — In spite of the apparent similarity, it must not be confused with the characteristics of a pressure relief valve. This similarity only comes from the fact that we are using the Z representation in which relief valves have horizontal characteristics and not the Y representation of fig 42—3.

movement of the main valve commences and leads to a much steeper increase. As before, the preload of the spring is responsible for the start of the movement and the spring constant for the inclination of the characteristics. Fig 4 contains, furthermore, the secant and the tangent to an operating point where the tangent practically coincides with the characteristic itself. They indicate the secant and tangent impedances as shown.

Good flow control valves, at favourable operating points (with maximum flow), have a tangent impedance about thirty times the secant impedance. In other words, the valve opposes a pressure increase of 30% to a flow increase of only 1%.In the Y representation, on the other hand, a change of pressure drop of 30% produces a change of flow of only 1%.

There are many different types derived from the basic form shown here, some of them with the main valve arranged sideways to the throttle valve. The throttle valve sometimes comprises a part sensitive to the change of temperature, compensating the change of viscosity by a change of the flow cross-section, so that the pressure drop with given flow remains approximately independent of the temperature. /FN$^3$/

It should be noted that flow control valves operate only with the represented direction of flow.Applying a flow in the reverse direction, the main valve goes into the wide-open position and only the impedance of the throttle valve remains.

Summarising dipole valves, their operation is represented by their characteristics between flow and pressure drop, as given by the tangent and secant impedances at each operating point. They can be classified as follows:

1. Laminar or viscous throttles with linear relation between pressure drop and flow, where tangent and secant impedances are equal and strongly depending on temperature.

2. Turbulent throttles with quadratic relation between pressure drop and flow. It follows from the properties of a parabola that the tangent impedance in each operating point is twice as large as the secant impedance. Both impedances are largely independent of temperature.

FN$^3$ — This compensation is not effective if the viscosity changes with given temperature, as with a different type of oil.

3. Pressure relief valves with very flat characteristics in Z representation where the secant impedance depends greatly and the tangent impedance only a little on the operating point. The tangent impedance is as small as possible compared to the secant impedance, down to about a ratio of 100.

4. Flow control valves with very steep characteristics in their operating range in the Z representation. /FN[4]/ The secant impedance again depends greatly, the tangent impedance a little, on the operating point. The ratio of tangent to secant impedance should be as high as possible and reaches values up to about 30.

FN[4] — Or with flat characteristics in the Y representation.

## 44    Directional Control Valves

Directional control valves are used to direct a flow into a hydraulic cylinder or actuator or to other consumers. Since the oil must be directed from the supply to one side of the piston and evacuated to the return from the other side, they are usually four-way valves. Directional control valves, therefore, connect both sides of an actuator piston alternately with the supply and the return conduit of the fluid.

Hydraulic directional control valves are all derived from the piston slide valves of the steam engines in the 19th century. Fig 1 represents one form with the supply port from above in the centre and with return conduits on both ports below. This type is also called a spool valve because of the form of the sliding valve body.

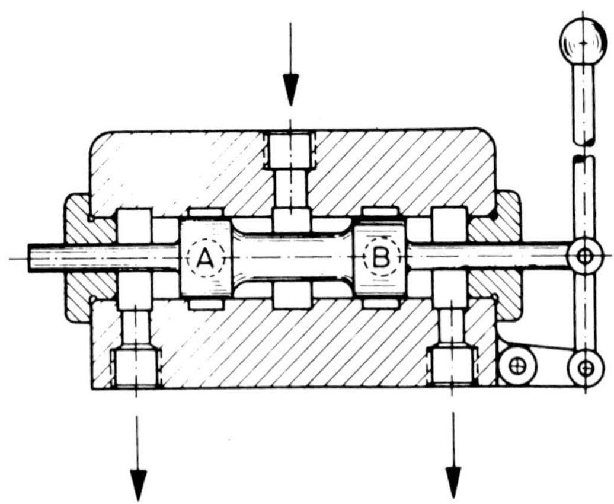

Fig 44.1: Schematic cross-section of a 4-way valve

The valve spool is shown, in fig 1, in the central position covering the cylinder ports denoted by A and B. All ports have the form of large annular grooves, in order to avoid any radially unsymmetric pressure fields, with the consequent danger of locking. Furthermore, the piston of the spool can be equipped with a number of small annular grooves, as discussed in Section 27.

The directional control valve in fig 1 operates as follows. Displacing the valve spool to the left connects port A with the supply and port B with the return port, whilst displacing it to the right connects port A with the return and B with the supply port.

In the central position of the valve spool in fig 1, all ports, and especially the cylinder ports, are blocked and this is, therefore, a *closed centre valve*. In other designs, called *open centre valves*, all ports are connected among themselves in the central position of the valve spool. In order to obtain this, the pistons of the spool in fig 1, are simply made shorter than the annular grooves of the cylinder ports A and B.

Starting from the basic model of fig 1, there are many models of directional control valves, actuated by hand, by electromagnets or, in the case of large models, by hydraulic piloting devices. The main spool of a piloted valve is displaced by hydraulic pressure which, in turn, is controlled by a smaller spool valve and a feedback arrangement. It thus represents a small servocontrol of the kind we shall deal with in Section 54. Due to the great number of different models, directional control valves are designated by two numbers, the first one giving the number of the ports and the second one the number of the possible positions of the valve body. Fig 1 contains, in this sense, a 5/3-way valve (since both return ports have separate external connections) and fig 2 a 4/3-way valve. The valve spool has, in both cases, three possible positions, ie right, centre and left.

For many applications, such as construction machinery, it is desirable to put several directional control valves in series, also called 'in tandem'. The oil should flow through the valve in the central position with a minimum of pressure drop and be directed to a cylinder or actuator only on displacement of the corresponding valve spool. For this purpose, the valve in fig 2 is used, having a passage from supply to return ports in the centre between both central pistons. This passage is blocked on moving the valve spool to the right or to the left when, at the same time, the cylinder ports A and B are connected with the supply and return ports. The compensation of hydrostatic longitudinal forces is important here, since considerable pressures may act in the return connections. It is obtained in fig 2, by the two outer pistons, which limit the circular spaces of the return ports in the axial direction, and have the same diameter as the other valve spool pistons.

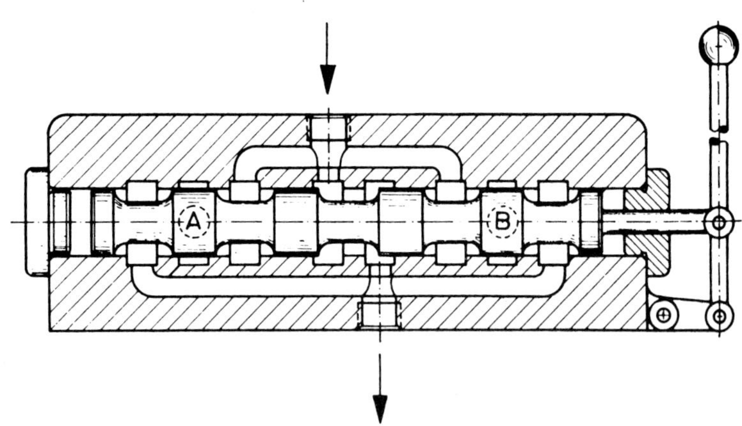

**Fig 44.2:Schematic cross-section of a 4-way valve with free passage of flow in the central position**

The hydrodynamic forces, or the forces connected with the inertia and the momentum of the oil jets, are given by our formula 41–1. These forces are proportional to the pressure drop on the metering edges between the valve spool and the cylinder ports and tend to close the valve. If several metering edges produce a distinct pressure drop, their forces must be added in order to find the net force on the valve spool.

Directional control valves are required to work satisfactorily only in certain positions, ie the central, the right and the left position of the valve body, and their performance in intermediate positions is much less important. In contrast, servovalves must also give satisfactory operation with partial openings.

Nevertheless, the relation between pressure drop and flow or the impedances and admittances in all positions are important. The pressure drop on a metering edge is governed by the inertia forces after our formula 13–8 with a loss coefficient $\zeta$ of about 2. Referred to the flow, this gives the following pressure drop

$$\Delta p = \zeta \, \frac{\rho}{2} \left[\frac{Q}{A}\right]^2 \qquad\qquad (44-1)$$

Equation 1 is applicable, according to B1, Section 7.1, if the Reynold's number is higher than 200 with the circumference of the groove as characteristic length and with the mean flow velocity. Furthermore, $\rho$ depends on small roundings or chamferings of the metering edges.

The pressure drop is proportional to the square of the mean velocity between the metering edges, or to the ratio between flow and cross section according to formula 13–8. The cross section itself is also given by the circumference of the metering edges and the stroke of the valve spool, leading to a large gain between opening and impedance. Especially when closing the valve, the impedance increases rapidly, leading sometimes to unpleasant shocks. The triangular (or Vee) grooves on the pistons of the valve spool in fig 3 produce a smoother closing by a less steep increase of impedance, since they change the flow cross-sections much more gradually with the movement of the valve spool. The impedances are again calculated by equation 1, using the cross-section of the triangular groove at the metering edge as a function of the valve position.

Fig 44.3: Valve spool with annular groove for radial
stabilisation and with triangular longitudinal grooves
for reducing impedance change on actuation

In order to avoid radial forces that would tend to press the valve spool against the wall, it is always necessary to provide two triangular grooves diametrically opposite or, in general, an even number of such grooves.

## 45   Servo valves

A servo valve is a directional control valve with electric input signal, giving with all positions of the valve spool, a determined relation between the electric control signal and the admittance. From the point of view of operation there are the following differences compared to normal directional control valves:

1. The admittance or the flow with a given pressure drop is proportional to the electric control signal.

2. There is no flow even with a large supply pressure without a control signal, but it begins if the control signal reaches only about 1.5% of its maximum value.

3. The servo valve can also follow very rapid variations of the electrical control signals, with a limiting or cut-off frequency around 30 Hz, depending on the model.

4. Even when wide open or with 100% control signal, the servo valve has an important pressure drop, usually of 70 bar (1 000 psi) with the rated flow of the servovalve.

The essential element of a servo valve is the valve spool, fitted as accurately as possible into the valve body or sleeve, in particular without axial underlap (opening) or overlap in the central position. As far as this is realized in practice, the flow area is exactly proportional to the stroke or displacement of the valve spool, with no lap travel from the central position. Also, the metering edges close immediately if the valve spool is displaced in the opposite direction, and the other edges open, supplying a reverse flow without threshold. The smooth transition between both directions of flow in the cylinder ports is important for the performance of servo controls and servo motors at low velocities.

The electric actuation normally uses a push-pull circuit. The magnetic coil has two windings, and its force is proportional to the difference of the electric currents in both.

The operation of an electrically controlled servo valve is easily studied on the functional diagram in fig 1. The input is the electric current i (or difference between the electric currents in both windings), producing the magnetic force. Together with the disturbing forces $F_{vs}$ it produces the stroke $s_{vs}$ of the valve body depending on the spring constant k, as indicated by the second block. The last block gives the relation between the position of the valve body, the pressure drop $p_{vs}$ and the flow Q.

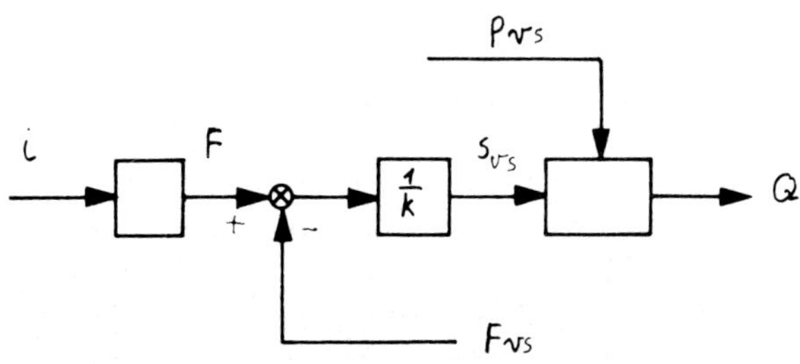


Fig 45.1: Functional diagram of a servovalve with direct electric actuation

The simple functional diagram in fig 1 requires the following comments:

1. The magnet produces a force that is transformed by a spring into the stroke of the valve spool, in turn determining the flow. A high gain, therefore, requires a soft spring.

2. The valve spool acts together with the spring as a mechanical resonator, limiting the response of the entire valve by its resonance frequency. As known from mechanics, a high resonance frequency is produced by a small mass and a stiff spring.

3. The force effects of the oil jets are additive to the magnetic force and their influence on the position of the valve body increases with a softer spring.

The spring is thus subject to contradictory requirements. It is necessary, since the magnet produces a force and not a position (stroke), as response to the electric current. In principle, other physical effects directly transforming the electric current into a stroke would be preferable, like the magnetostriction known from physics. Unfortunately, all known physical effects are either too weak or too inaccurate to be useful in oil hydraulics. (Private communication by Prof. S.Y.Lee, M.I.T., Cambridge, USA).

In order to reduce the problems with the spring, servo valves are normally equipped with a small piloting valve, or force amplifier, between the magnet and the main valve spool.

As an introduction fig 2 represents a valve spool as part of a servo valve supplied in the middle by a pump. The load is a hydrostatic motor, and the four metering edges are clearly visible. Two of them are closed depending on the position of the valve spool and the flow must pass one metering edge each time, when going or returning from the motor.

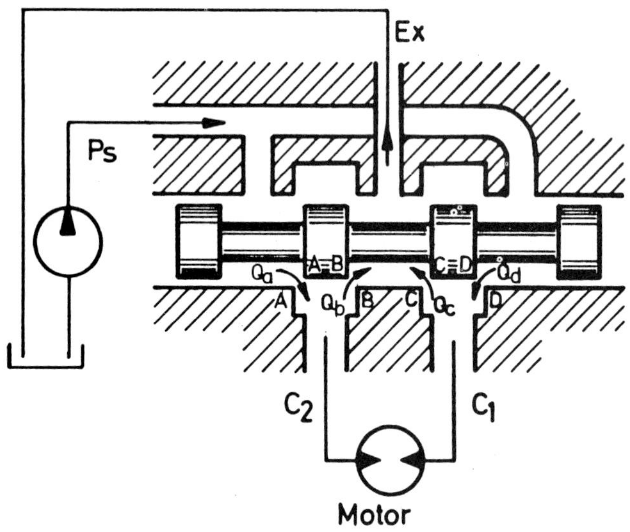


**Fig 45.2:** Valve spool of a piloted servovalve (Intramat)

Fig 3 represents a cross section through an electro-hydraulic servo valve, with the magnet and the electric connection 1; on the top the coil 2 and the moving piece. Its lower prolongation carries the flapper 7 which extends between the two jets 9 into a recess of the valve spool 8. The jets are fed over two throttles 10 and are in parallel with the end faces of the valve spool, whilst a small filter is fitted in the supply duct for the protection of the sensitive throttles. The spring tube 6 positions the movable piece and the flapper adjustable in the casing and, at the same time, seals the space of the valve spool against the space of the magnet. This sealing is important to prevent any iron particles in the oil from getting attached on the magnetized parts. This effect is similar to magnetic filters, but here produces sticking perturbations.

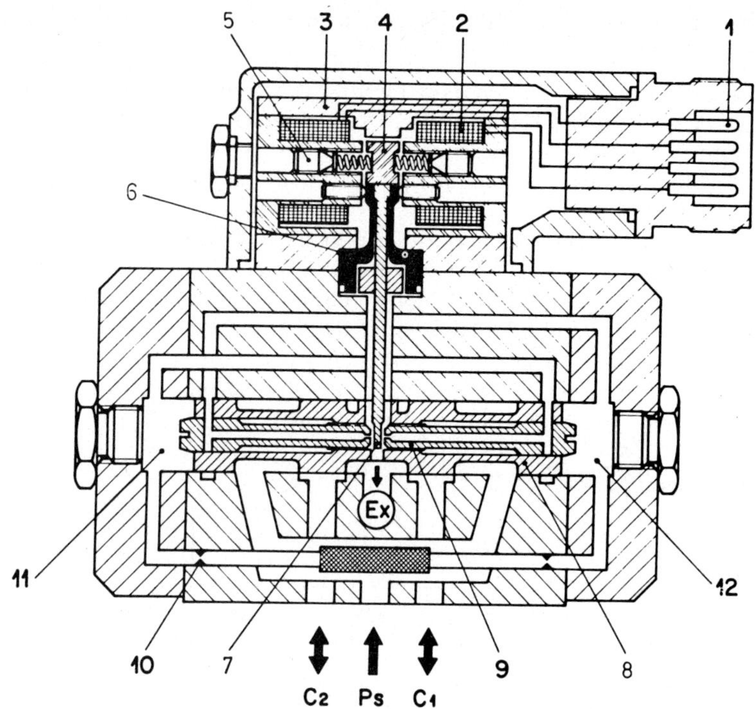


Fig 45.3: Cross-section through an electric piloted
servovalve with flapper and two plates (Intramat)

The moving piece with the flapper is also located by the small
coil spring and the screw 5. This screw allows an adjustment of the
neutral (no electric current) position of the flapper and of the valve
spool. The spring effect, transforming the force of the magnet into
its position, is obtained by the collaboration of these two coil
springs and the spring tube.

The valve is supplied by the lower connection $p_s$. The loads are
connected over the ports C1 and C2 and the oil returns over the
connection $E_x$ in the centre near to the valve spool. For a correct
operation of the flapper, the jet 9 on the right is connected over the
duct with the left end face 11, and with the left throttle 10, similarly,
the left jet with the right end face of the valve spool and the right
throttle. If the flapper moves to the right, a larger pressure drop is
produced on the nearer jet end at the left end face, driving the valve

spool to the right, until the flapper is again centrally between both jets.

The piloting device consists thus of two fixed throttles and two flappers as variable impedances. It has no friction and the force effects of the oil jets compensate each other in the central position. Fig 4 obtained from fig 1 by adding the third block on top, shows the operation as a functional diagram. This block gives the relation of the gain between the position of the flapper and the valve spool. It is equal to one, as shown with low frequencies, whilst the inevitable delays must be considered at higher frequency. The important fact is that the dynamic and other forces on the main valve spool have only little effect back to the flapper, as shown by the lower block with vanishing gain. This allows us to use a soft spring without disturbances of the flapper position by these forces.

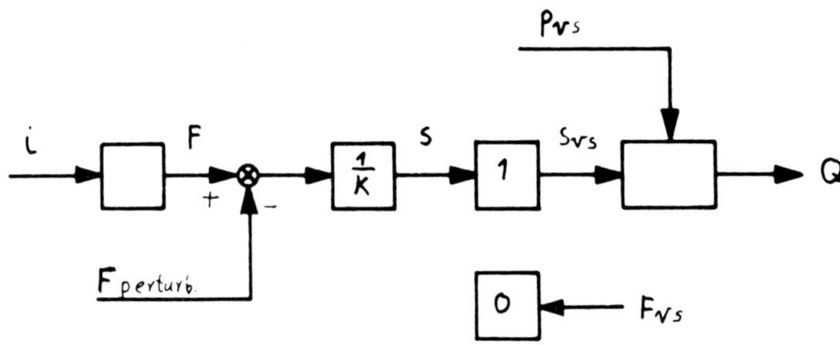

Fig 45.4: Functional diagram for a piloted servovalve

The position of the valve spool is proportional to the electric control signal, and the flow is connected with the valve pressure $p_{vs}$ over the turbulent pressure drop at the metering edges. Since there are always two metering edges in series, we obtain with $p_{vs}$ denoting the entire pressure drop:

$$Q = b\, A_{vs} \sqrt{\frac{p_{vs}}{\zeta \rho}} = b \sqrt{\frac{p_{vs}}{\zeta \rho}}\; A_{max}\, \frac{i}{i_{max}} \qquad (45-1)$$

where the last form immediately gives the relation between electric current and the position of the valve spool. The flow increases linearly with the electric current, but still depends largely on the pressure drop.

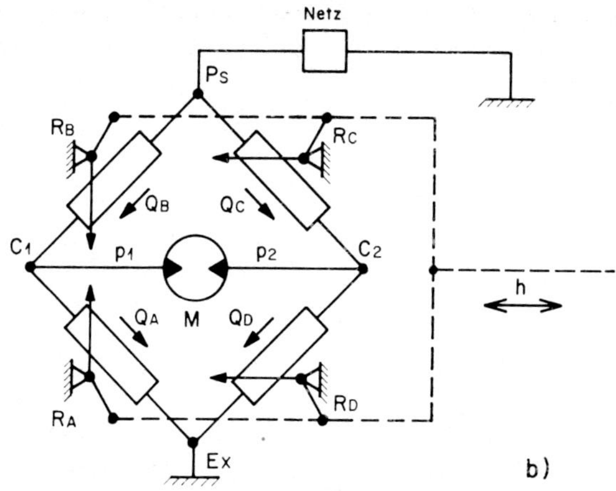


**Fig 45.5:** Representation of the operation of a servovalve as bridge circuit (Intramat)

The valve spool has 4 metering edges, and is represented in fig 5
as a bridge circuit with a hydraulic motor as load. The pressure at
the motor is connected as follows with the supply pressure $p_{sp}$:

$$p_{ld} = p_{sp} - p_{vs} \qquad\qquad (45\text{--}2)$$

Usually one represents the relation between the flow and the
pressure across the load $p_{ld}$ and not the pressure drop $p_{vs}$, in the
function of the electric control signal. This is shown in fig 6.
Examining first the upper part, there is no flow if the pressure on the
load $p_{sd}$ is equal to the supply pressure $p_{ld}$. With smaller load
pressure, we have the turbulent characteristics according to equation
1, associating a positive flow to a movement to the right, and a
positive pressure drop on the load to the usual case where the load
opposes this flow. Any force in the other direction, as when the load
is braked by the servovalve, is described by the continuation of the
characteristics on top left until a negative $p_{ld}$.

Reversing the electric control current, the spool valve connects
the other cylinder port with the supply, leading to the image characteristics on the lower half of fig 6. A negative flow normally produces

a negative pressure drop, below at left, and only exceptionally (if the load is reverse braked by the servovalve), a positive pressure drop, represented below at right.

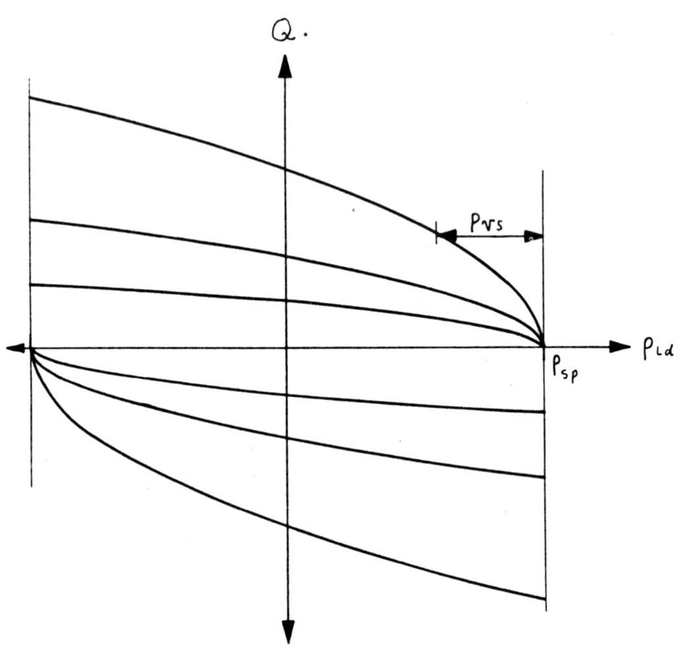


Fig 45.6: Characteristics of a servovalve

Servovalves of this or a similar kind have been developed for aircraft and missile control systems, and are used today (1970), for many control problems. They have rated flows of 5 up to 500 lit/min under full control signal with 70 bar pressure drop (1 to 80 gpm at 1 000 psi), with the most usual value of about 40 lit/min (10 gpm). The consumption of the piloting device is between 1 to 2 lit/min with 70 bar supply pressure and increases with its root. The reader is referred to Guillon B7 for a more detailed treatment of this interesting component.

## 46    Other Components

There are many other components of oil hydraulics, but here we shall only briefly discuss the most interesting ones.

### 46–1    Hydraulic Accumulators

In many hydraulic installations oil under pressure and thereby hydrostatic energy should be conserved in an accumulator until needed again. Basically there are two different applications:

1. To have hydrostatic power available in emergencies or for a short peak power demand.
2. To absorb kinetic energy liberated on change of the operating condition, producing otherwise undesirable shocks or pressure peaks.

Accumulators therefore keep potential energy for some time /FN[1]/ Generally, potential energy can be kept by any matter or material under stress, and also by lifting weights. Accumulators with a weight loaded piston are no longer used since they are too large and too expensive. The disadvantage of accumulators with elastic stress as in springs is the very small admissible strain or deformation. The strain of metals at the elastic limit, the highest limit for the admissible stress, is only about (3% o (3/10%). Metals can, therefore, accumulate only very little energy compared to their mass, in spite of high stresses. They are also subject to fatigue failures, especially if they are frequently and rapidly loaded and unloaded.

Gases have the advantage of being deformable indefinitely according to the so-called ideal gas laws, without ever being damaged. They admit the usual pressures of oil hydraulics without difficulty and their mass is always small compared to the mass of the container.

FN[1] — The flywheel constitutes an analogous mechanical accumulator, and keeps kinetic energy by its movement, but the losses are much higher if the energy has to be kept for any length of time, due to friction etc.

With these properties gases are the most advantageous accumulators for energy, but they must be strictly separated from the hydraulic oil, in order to avoid diffusion or mixing. The separation can be achieved as follows:

1.  Separation by a fitted piston

2.  Separation by a deformable wall of elastomer

There are many industrially used piston accumulators and also many accumulators with a deformable wall made from elastomer. The wall either has the form of a membrane in a spherical container, or the form of a bladder suspended only at the gas entrance connection.

Fig 1 represents a bladder-type accumulator, with the entrance for the oil below, and the gas connection with suspended bladder on top. The operation of the accumulator depends on the filling pressure of the bladder, and even with perfect sealing there are certain losses of gas by diffusion through the walls, so that the gas pressure must be topped up once every twelve months. In order to avoid extrusion of the bladder through the fluid connection, with consequent damage, there is a non-return valve that closes on contact with the bladder. The layout must be such as to prevent the bladder being squeezed by

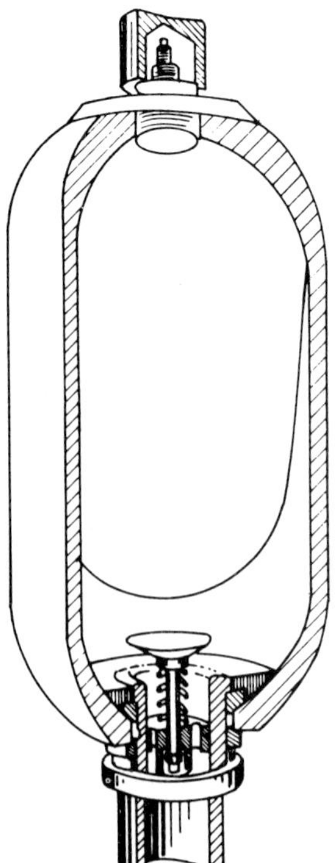

Fig 46.1: Schematic cross-section through
a gas-loaded hydraulic accumulator

the valve seat. Instead of the valve, sometimes a sheet metal with well-rounded perforations, is used.

Due to the danger of fire (resembling an explosion under high pressure) the accumulator must be filled only with non-flammable gas. Normally one uses nitrogen, although the noble gas helium would have more favourable properties, such as less chemical attack of the wall. A great advantage of gas accumulators is that their characteristics can be influenced by the gas filling pressure. They accept hydraulic oil only if this pressure is exceeded, corresponding to the volume reduction of the gas according to the ideal gas laws. The accumulated mechanical energy appears as heat, again absorbed by the gas when the accumulator is unloaded. Depending on the heat capacity, the temperature increases on loading and decreases on unloading of the accumulator, with some influence on the characteristics.

There are two limiting cases for the characteristics:

1. Isothermal characteristic with constant gas temperature.
2. Adiabatic characteristic, where the heat is not absorbed by the walls, but entirely transformed into temperature increase of the gas.

The characteristics between the accumulated volume V and the applied pressure p is in the isothermal case:

$$V = 0 \qquad \text{if } p < p_{fill}$$

$$V = V_0 \left[ 1 - \frac{p_{fill}}{p} \right] \quad \text{if } p > p_{fill} \tag{46-1}$$

with $V$ = volume of the accumulator, $p_{fill}$ = gas filling pressure,

and in the adiabatic case:

$$V = V_0 \left[ 1 - \frac{p_{fill}}{p}^{k} \right] \text{if } p > p_{fill} \tag{46-2}$$

$$\kappa \approx 1.4$$

According to B2, page 21, the real characteristic approximates closely to the adiabatic case. According to the same source, in normal service only 60% of the volume $V_0$ should be emptied and filled at each cycle and for a long life with very frequent cycles, even only 40%.

Accumulators of this kind are an interesting component. Their usefulness is not limited to store energy, but they can eliminate pressure peaks and vibrations, as mentioned above.

## 46-2    Filters

We have already noted in Chapter 2 the damaging effect of contamination and impurities in the oil, in the fluid mechanics of the gaps. Hydraulic components with low gap heights, having in principle the advantage of low leakage, are especially sensitive to small solids suspended in the oil. In a certain sense, this contamination sensitivity is the main problem of oil hydraulics.

Although the hydraulic oil is not contaminated by the combustion residues as in internal combustion engines, impurities are nevertheless continuously produced by the following sources:

1. Particles broken from the walls during normal wear.
2. Penetration of dust particles during operation.
3. Remainders from manufacture and maintenance procedure.

Real contamination does not consist of particles of a certain size or diameter, but the number of particles increases continuously with decreasing diameter. In order to arrive at a quantitative indication, contamination classes have been set up. Table 1 contains the contamination classes established by NASA 1638 (USA) with 14 contamination classes beginning with the cleanest class 00 to the most dirty class 12. In industrial installations one mainly finds classes 10 and above.

The admissible number of particles increases in each class with decreasing diameter, but even in the very clean classes a few large particles are tolerated.

It is necessary to control the contamination of the oil by filters, and also new hydraulic oil is usually supplied with too many impurities. Therefore, practically all hydraulic installations are equipped with filters, except perhaps some with large gap heights.

The filtration itself can take place according to one of the two following principles:

1. Filtration by sedimentation.
2. Filtration by mechanical straining.

### Sedimentation Filters

In sedimentation filters the particles are removed by suitable forces from the fluid and deposited where they can do no harm. The simplest example is a large oil tank, which can be very efficient if the sedimentation is not disturbed by convection currents or

## Table 46–1 NASA Contamination Classes

Admissible Number of particles per 100 cm³

| Size Range (Micro-meter) | Classes No | | | | | | | | | | | | | |
|---|---|---|---|---|---|---|---|---|---|---|---|---|---|---|
| | 00 | 0 | 1 | 2 | 3 | 4 | 5 | 6 | 7 | 8 | 9 | 10 | 11 | 12 |
| 5 – 15 | 125 | 250 | 500 | 1 000 | 2 000 | 4 000 | 8 000 | 16 000 | 32 000 | 64 000 | 128 000 | 256 000 | 512 000 | 1 024 000 |
| 15 – 25 | 22 | 44 | 89 | 178 | 356 | 712 | 1 425 | 2 850 | 5 700 | 11 400 | 22 800 | 45 600 | 91 200 | 182 400 |
| 25 – 50 | 4 | 8 | 16 | 32 | 63 | 126 | 253 | 506 | 1 012 | 2 025 | 4 050 | 8 100 | 16 200 | 32 400 |
| 50 – 100 | 1 | 2 | 3 | 6 | 11 | 22 | 45 | 90 | 180 | 360 | 720 | 1 440 | 2 880 | 5 760 |
| Over 100 | 0 | 0 | 1 | 1 | 2 | 4 | 8 | 16 | 32 | 64 | 128 | 256 | 512 | 1 024 |
| Total No. | 150 | 300 | 600 | 1 200 | 2 400 | 4 800 | 9 700 | 19 400 | 38 900 | 77 800 | 155 000 | 311 000 | 622 000 | 1 245 000 |

larger movements of the fluid like caused by temperature differences. It is also necessary to arrange the return oil conduit in a way that the flow does not stir up the oil. The sedimentation force in the tank is simply the weight of the particles in the oil, diminished by their floatation force.

In industry the size of the oil tanks is usually selected so that its volume can take the highest flow during a time of two minutes. If bulk and mass must be economised, the size may be reduced to about one minute.

In centrifugal filters, the sedimentation is assisted by the centrifugal forces. An interesting design uses a rotor, driven by a small jet turbine, fed by the oil flow itself. This avoids a mechanical drive shaft and the pressure drop of the jet turbine does not normally disturb. It should also be noted that both in the centrifugal filter and in the tank, the sedimentation force is proportional to the difference between the mass density of the particles and of the fluid. It is therefore not very efficient for particles of a similar mass density as the oil.

In magnetic filters the sedimentation is assisted by magnetic forces, acting in principle only on magnetizable materials, like iron, steel and $Fe_3O_4$. In practice, however, most of the impurities contain steel particles, since in the case of abrasive wear, bronze parts break out together with steel particles, and magnetic filters are surprisingly efficient. This includes fibres, usually impregnated with steel particles. For an efficient action, vortices and convection currents should be avoided in the magnetic field, as in the oil tank. Furthemore, the sedimentation forces are proportional to the gradient of the magnetic field and therefore in a strong but homogenous field, no force is acting. Magnetic filters of this kind have the advantage that they also eliminate very small particles below $2\mu$m diameter.

A general advantage of sedimentation filters is their large dirt holding capacity, permitting maintenance at infrequent intervals.

*Mechanical Filters*

Filters that eliminate particles mechanically by a strainer are made in the following basic types:

1. Surface filters
2. Volume filters

The surface filters contain the strainers or sieves used with openings down to 50 $\mu$m. For finer filtrations impregnated papers or proprietary materials are used. The filtrating surfaces are often arranged in a star-shape, in order to accommodate a large area in a given size casing.

With very fine filters, the retained impurities are quickly accumulated in front of the openings, leading to an increased pressure drop but without impairing the fine filtration. In order to avoid rupturing of the filter surface under too high a pressure drop, a relief valve is usually put in parallel. Naturally, if it opens, the flow will no longer be cleaned. A recent tendency are filters giving an alarm signal with too much dirt, actuated by the increasing pressure drop. The signal can take the form of an indicator going into a red area or of a pilot lamp, actuated by a pressure switch.

In the volume filter the oil flows through a thick filter cartridge, supposedly catching the impurities. They are preferably used for small flows.

In order to combine the advantages of magnetic and mechanical filters, it is possible to arrange both principles in a common casing as shown in fig 2. We find the inlet connection at right, directing the oil to a magnetic assembly or magnetic candle in the axis. It is composed by magnets of high coercive force, usually of unisotropic barium oxide, leading to the production of strongly inhomogeneous fields. The flow path is designed for low vortex generation as described above.

Fig 46.2: Combined mechanical and magnetic filter with central magnetic candle and strongly inhomogeneous magnetic field (Regeltechnik)

The oil in fig 2 flows subsequently from the inside to the filtering surface, retained at the outside by the visible perforated sheet metal, and finally reaches the outlet at left. Filters of this kind are built for flows from 5 to 500 lit/min (1 to 100 gpm).

It should be noted that formerly two filtration ratings were used, viz:
1. The nominal filtration ratings, on which 98% of the mass of all particles with diameters below, should be eliminated.
2. The absolute filtration rating, where no particles with larger diameter are supposed to pass.

The actual determination of the contamination is uncertain and demands special expensive instruments. It is nevertheless necessary, in order to obtain satisfactory results with hydrostatic components having small gap heights. Finally, for a more exhaustive treatment of filtration and contamination assessment, the reader is referred to the work of Prof. Fitch, of Oklahoma State University, USA.

The present (1970) tendency is to use only the absolute filtration rating, or better still the contamination classes on Table 1. With a given filter, filtration ratings were usually apart by a factor of 2, meaning that a filter with a nominal rating of 10 micrometer had an absolute filtration rating of 20 micrometer.

# 5   HYDROSTATIC INSTALLATIONS

Hydrostatic components are not used singly, but combined in installations in order to obtain a desired effect or service. Hydrostatic transmissions are an especially important combination, with essentially two hydrostatic machines, and will be treated in chapter 6. Here we shall examine other hydrostatic installations which we shall classify as follows:

1.   Hydraulic power packs

2.   Hydraulic handling

3.   Hydraulic presses

4.   Servomechanisms

### 51      Hydraulic Power Packs

The hydrostatic energy is produced in hydraulic power packs, comprising basically an oil tank with a pump, usually electrically driven, and some valves and filters.

The simplest power pack delivers a constant flow up to a maximum pressure, set by a pressure relief valve.

Fig 1 represents the circuit of a small hydraulic power pack, with a pump with fixed displacement, left. The maximum pressure is limited by the relief valve in the centre and the outflow returns over a filter to the tank. In normal operation, a part of the flow of the pump goes through the relief valve, keeping a constant pressure available at the output connection at right. The relief valve destroys a power given by the product of its flow with the operating pressure and transforms it into heat.

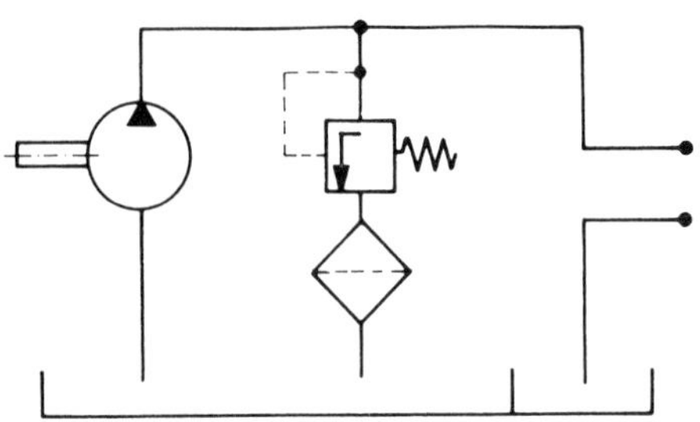

Fig 51.1: Circuit of a power pack with constant displacement machine

For the dimensioning of a power pack it should be noted, that the maximum pressure determines the available force together with the area of the actuator. The required speed of advance of the cylinder fixes with its area the required flow.

In order to avoid the power loss in the relief valve it is advisable, especially with larger powers, to build a power pack with a variable displacement machine, where the flow of the pump is controlled either by a pressure or by a power regulator.

## Pressure Regulators

Pressure regulators are used to adjust the displacement of a variable pump so that the same pressure is maintained in the delivery tube independent of the variations of the flow demand. Naturally this is possible only until a maximum or rated flow. Such regulators are built essentially in two different designs:

1.  Pressure regulators with a cylinder and a spring

2.  Pressure regulators with a separate control circuit

## Pressure regulators with cylinder and spring

Fig 2 represents the circuit of a pressure regulator with cylinder and spring. In contrast to fig 1, the operating pressure is conducted over a throttle and a non-return valve in parallel with the actuating cylinder, thus producing a force. The displacement of the volume is reduced by the actuator as soon as this force exceeds the preload of the spring. Conversely, the spring increases the displacement again if the pressure falls. The non-return valve in the pressure line to the actuator admits a rapid reduction of the displacement and of the flow delivered by the pump. With increasing pressure, the speed of the actuator increasing the pump displacement is limited by the throttle, improving the stability of this circuit.

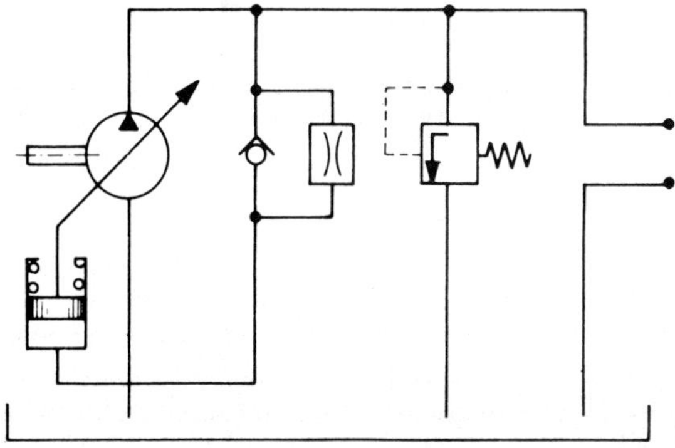

Fig 51.2: Circuit of a power pack with adjustable displacement and pressure regulator

Fig 3 represents the characteristics of a power pack with pressure regulator in the Y representation. The pressure regulator should have a steep characteristic in its operating range, or a large tangent admittance for the flow, ideally infinitely large. Furthermore it is desirable that the set value of the pressure can be easily adjusted.

The pressure regulator with cylinder and spring has the following limitations:

1. The spring must be very soft, have a considerable preload and generally produce forces large enough for a secure displacement control.

2.  It is difficult, from the design point of view, to provide an adjustable preload of the spring for adjustable set pressure.

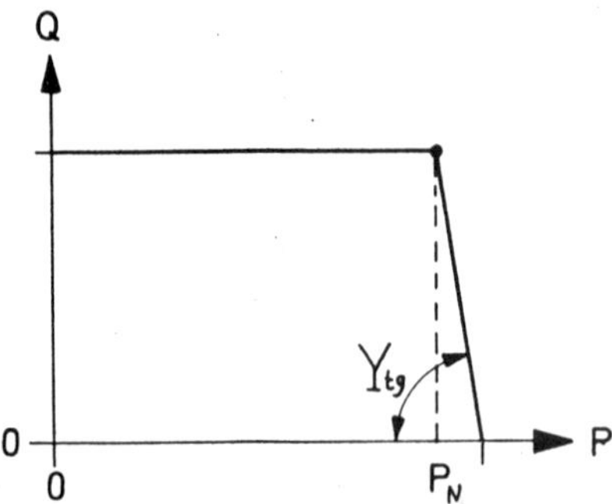

**Fig. 51.3: Characteristics of a hydrostatic machine with pressure regulator**

The actuating cylinder is, together with its spring, a kind of pressure gauge acting directly on the displacement setting. In order to avoid disturbing forces it is often preferable to interpose a servomotor or force amplifier, similar to servovalves. Furthermore all such installations should be equipped with a separate pressure relief valve in order to avoid dangerous pressure peaks, especially with rapidly increasing loads.


## Pressure regulator with separate control circuit


In order to avoid the difficulties of the highly loaded spring in the actuator, we can provide a separate control circuit as shown in fig 4. The delivery tube is connected to a pressure relief valve, taking out a certain flow on reaching its set value. This flow enters the actuating cylinder, containing a weak spring, and in parallel with the throttle. The pressure in the actuating cylinder vanishes as long as the pressure relief valve is closed and only when it opens, the throttle produces a pressure drop that enters the actuating cylinder in order to reduce the displacement of the pump. A great advantage is that the actuating pressure $p_{ac}$ can be freely selected normally in the range of 25 to 30 bar (400 to 600 psi), leading to a simple and light spring.

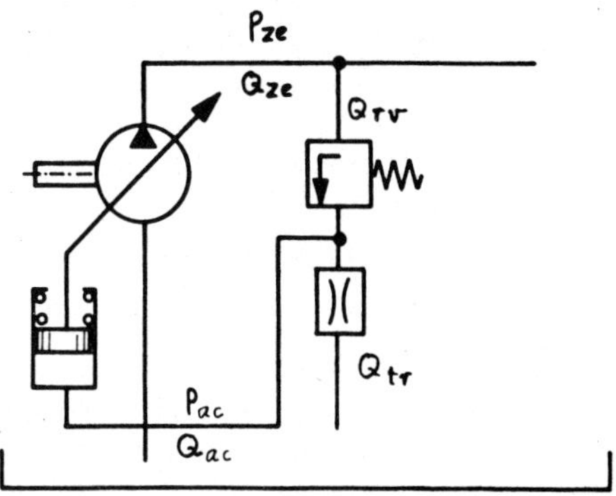


Fig. 51.4: Circuit of a pressure regulator with separate
control circuit and relief valve

We want to study the operation of this pressure regulator by the
functional diagram in fig 5. It contains:

1. The delivery pressure determines the flow $Q_{rv}$ by the admittance $Y_{rv}$.

2. The flow $Q_{tr}$ produces together with the impedance of the throttle Z the actuating pressure $p_{ac}$.

3. The actuating pressure gives the force $F_{ac}$ by the area of the actuating piston $A_{ac}$, and furthermore the position of the actuating piston by the spring constant k.

4. The position of the actuating piston controls the displacement setting of the pump and thereby the flow $Q_{ze}$.

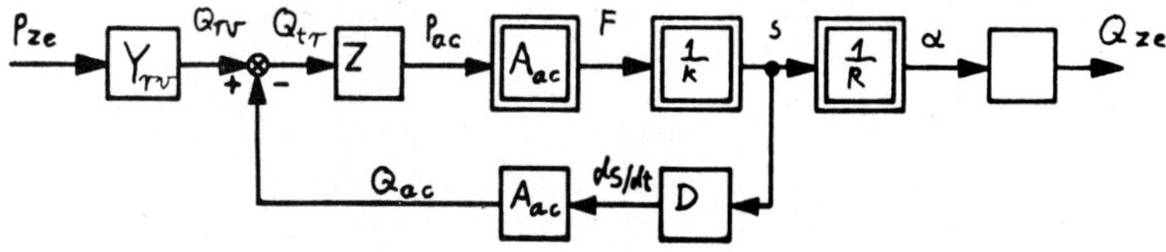


Fig. 51.5: Functional diagram of the pressure regulator
with relief valve and separate control circuit

So far we have already described quantitatively the static operation of the pressure regulator. The lower line of fig 5 is only needed to study very rapid adjusting movements.

5. From the position of the actuating piston we determine the velocity by derivation with respect to time, as represented by the lower block right with the operator D.

6. Multiplication with the area of the actuating piston $A_{ac}$ gives the flow in the cylinder $Q_{ac}$.

7. This flow is deducted from the flow of the pressure relief valve, as indicated by the subtraction point.

With sinking delivery pressure, the relief valve closes and the spring again increases the displacement of the pump. The velocity of of this reset is determined by the spring force, by the area of the control piston and by the pressure drop in the throttle. It is more conveniently studied by inversion of the functional diagram in fig 5, but will not be pursued here.

Fig 6 contains a view of a hydraulic power pack of this kind. It is equipped with an electrical drive motor with vertical shaft, a variable displacement axial piston machine within the oil tank, with several valves and a pressure gauge.

**Fig 51.6: Hydraulic power pack (Galdabini)**

## Power Regulators

The word 'power regulator' is an unfortunate choice, but generally accepted. It really denotes a control of the displacement volume and of the flow in function of the operating pressure so that the product of these two variables remains approximately constant. In this case the hydrostatic machine also needs a constant torque by our fundamental equation 32—2, corresponding to a constant power if the rotation frequency remains constant.

The main differences compared to the pressure regulator are the curve characteristics. In principle they should approximate to hyperbolas according to the following equation

$$\alpha = \frac{M_1}{q_o} \frac{1}{p} \tag{51—1}$$

where $M_1$ denotes the desired input torque.

The hyperbola is frequently approximated in practice by broken lines as shown in fig 7 in the Y representation. This broken line is built simply by putting two or three springs of different length in parallel, preferably within the control cylinder. In contrast with the

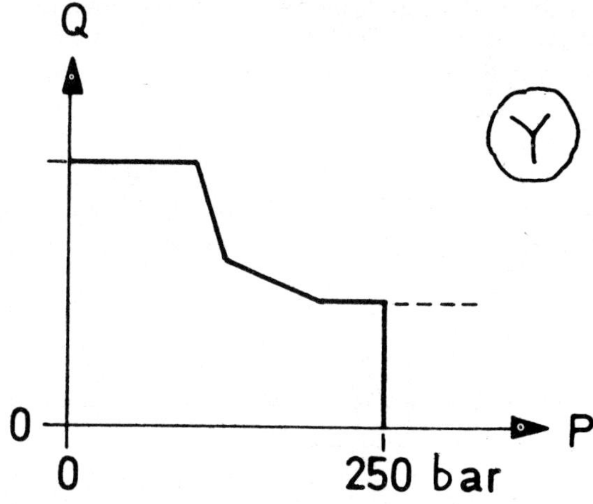


Fig 51.7: Characteristics of a hydrostatic machine
with power regulator

pressure regulator, here the springs have a certain slope of characteristic and are not so soft. Under sufficiently high pressure, the displacement volume remains constant, as shown on the horizontal part of the characteristic in fig 7 at right. Finally with a maximum pressure 250 bar (3 800 psi) the flow is diverted by a separate pressure relief valve. This part of the characteristic is not produced by the power regulator itself and without the relief valve the characteristics would continue to higher pressures with consequent damage.

Power regulators of this kind are frequently used in hydraulic presses, direct acting in small sizes up to flows of 150 lit/min and with a piloting device (force amplifier) for larger flows.

The stability is an important question, since a hydraulic power pack with such a regulator can produce oscillations similar to a transistor. The basic reason is, that the power pack can supply a hydrostatic ac (alternating current) power, caused by fluctuating flow and pressure, whilst the power needed to change the displacement setting is much smaller. The difference between both powers is supplied by the drive shaft, in the same way as the power leads of the transistor.

The stability is best examined by the functional diagram in fig 8. Its input is the set pressure and the first block represents the action of the regulator, as was described more in detail by fig 5. It controls the displacement setting of the pump and, by the second block, the flow delivered. This flow produces the operating pressure over the impedance of the load, as shown by the lower block. The operating pressure acts on the regulator over the shown subtraction point.

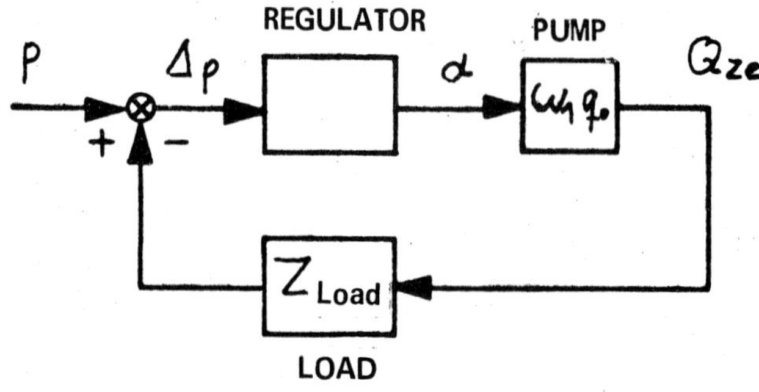


**Fig. 51.8: Functional diagram for the stability of pressure regulator**

The subtraction point is realised either by the opposing effects of piston and spring in the cylinder, or by the pressure relief valve in fig 4.

The functional diagram in fig 8 can easily become unstable, producing undesirable oscillations of pressure and flow. These questions can be examined by the methods of control engineering, together with ways for improving the stability, such as suitable feedback arrangements and others. Here we shall only note that the stability in the functional diagram becomes less favourable the more the loop gain is increased. A part of the gain of the loop in fig 8 is the relation between pressure and displacement setting. This gain must be increased if the regulator is required to work more accurately, that is to produce a steeper characteristic in fig 3, thereby leading to a higher loop gain and to less stability margin.

Pressure regulators are required to have especially steep characteristics (in the Y representation) and are therefore more liable to instability. Power controls have less inclined characteristics, as shown in fig 7, and tend to be more stable. Finally it should be noted, that the stability depends on the impedance of the load as part of the loop gain in fig 8. A power pack with pressure regulator must therefore have a sufficient stability margin, if it is to work satisfactorily with different loads or consumers.

## 52      Hydraulic Handling

Hydraulic handling frequently calls for movements which require large forces by actuating cylinders. Together with hydrostatic transmissions, it is the most important application of oil hydraulics. By a suitable combination of cylinders and valves one can produce a large variety of different movements or movements in different sequences. This leads to many different circuits, all composed from some basic elements, as discussed in this section.

Fig 1 shows a simple handling installation with a directional control valve in an actuating cylinder. The pressure oil is taken from one of the power packs of section 51 over the connections in the centre of the valve at left. This valve, represented by its conventional symbol, is a simple directional control valve as described in section 44. The actuating cylinder moves upwards or downwards depending on the position of the valve spool.

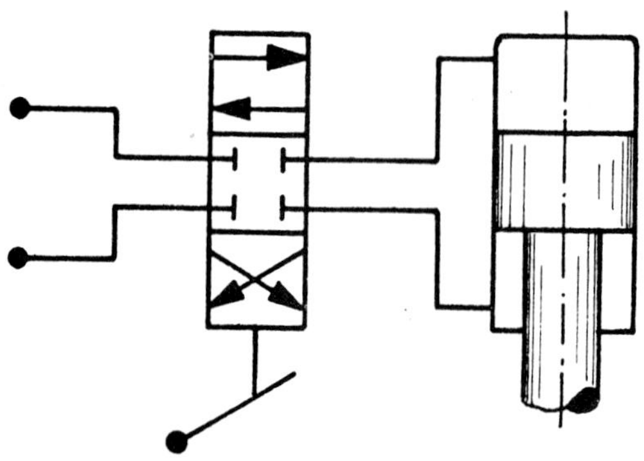

Fig 52.1: Circuit of simple hydraulic handling

We have the following simple equations for the relation between piston velocity and flow on the one hand, and between force and pressure on the other hand, with A = piston area

$$v = \frac{Q}{A} \qquad\qquad F = pA \qquad\qquad (52\text{--}1)$$

In spite of the simplicity of the circuit in fig 1 the following points should be noted:

1. The piston rod has a large diameter, occupying an appreciable part of the cylinder. The effective area of the piston is therefore smaller on return, leading to a higher return velocity with given flow. According to equation 1, less force is produced by a given pressure, but in many cases this is not important.

2. Fig 1 contains a directional control valve with blocked flow in the central position, thereby blocking the piston in the actuating cylinder. If a freely movable piston in the centre position of the valve is desired, a control valve with open ports, called an *open centre valve*, can be used.

Fig 2 represents a slightly more involved circuit with two cylinders. The cylinders are connected in parallel, and the circuit has the following notable features:

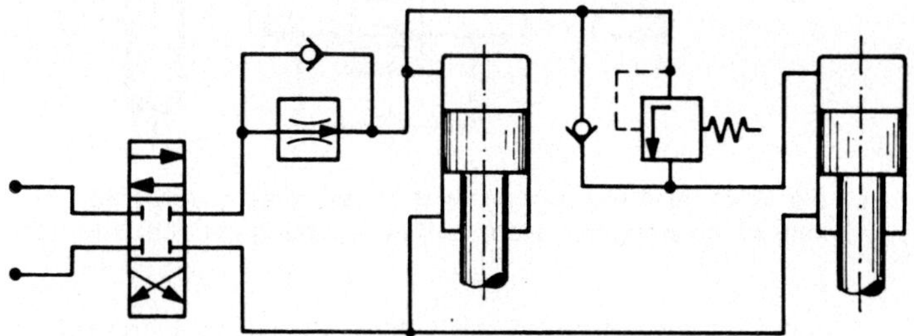

**Fig 52.2:** Handling circuit with two cylinders and sequence valve

1. The flow control valve in the forward connection, bridged by a non-return valve, limits the flow and the velocity of the piston, especially important in large installations. This circuit is also called metering in flow or speed control.

2. The second cylinder is connected through a sequence valve bridged by a non-return valve. A sequence valve is simply a pressure relief valve admitting flow to the second cylinder at

right only upon reaching a set pressure. The second cylinder
starts its movement therefore only after the first cylinder
encounters a resistance and thereby increases the system
pressure. The circuit in fig 2 thus assures a certain sequence
of operation. In a typical application, cylinder 1 would fix or
clamp a work piece, and cylinder 2 initiate a machining operation, only after the work piece is securely held by cylinder 1.
Setting the directional control valve to reverse, the pistons
return rapidly since the flow goes through both non-return
valves.

Fig 3 shows another handling arrangement with a flow control
valve in the return connection, also called metering out speed control. Consequently, the actuating cylinder contains high pressures,
improving the stiffness of actuation specially if slight air inclusion
makes the oil more compressible.

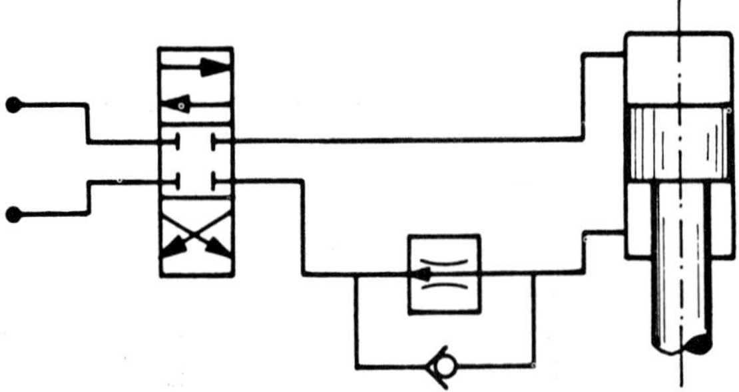

**Fig 52.3:** Handling  circuit  with  flow  control  valve  in
return  connection,  also  called  metering-out  circuit

Fig 4 contains an actuation with three directional control valves
in series, namely two actuating cylinders and a hydrostatic motor.
The directional control valve allows the flow to pass when in the
central position with a minimum of pressure drop, using a design as
shown in fig 44—3.

The series circuit has special advantages if normally only one
output is required at a time. It is frequently used on construction
machinery, where up to 10 directional control valves, arranged in
a bank, are connected in series and fed by a flow source. The pressure drop in the central position of each directional control valve
should be as small as possible to avoid too high losses in the entire

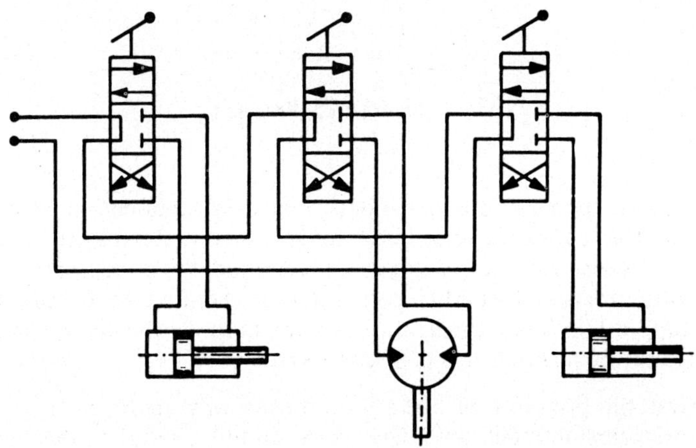

**Fig 52.4: Handling circuit with three loads in series**

bank of valves. Other handling installations are completed by further flow control or pressure relief valves, depending on the requirements. These auxiliary valves are frequently placed inside the casing of the directional control valves.

## 53      Hydraulic Presses

Hydraulic presses are the oldest industrial application of hydro-statics. They produce very large forces, mainly for metal shaping, by a cylinder under fluid pressure supplied by a separate pump. Hydraulic presses formerly used water as working fluid, but today hydraulic oil is more frequent. They are built in the force range of 5 kN up to 100 MN (0.5 to 1 000 tons force).

Whilst the principle of hydraulic presses was shown in the figure |in the historical\introduction, they use a circuit similar to that shown in fig 52—1. Smaller presses are normally equipped with a power pack with a fixed displacement pump and a circuit similar to fig 51—1. Larger presses preferably incorporate hydrostatic machines with an adjustable displacement and power regulator as shown in the circuit of fig 51—2, leading to an important saving in power requirements. Both with fixed and adjustable displacement *open circuits* are frequently used. In the open circuit, the oil is pumped from the tank over a directional control valve into the press cylinder and then returns to the tank.

Many hydraulic presses are equipped with several pistons. As an example a metal sheet may be blocked first by a small piston, before the large press piston begins its work of deformation, normally using the sequence circuit of fig 52—2.

A rapid return stroke is especially desirable with hydraulic presses, since only the weight of the (usually vertical) press piston must be overcome. Thus the return piston area, that is the area of the lower face of the piston, is smaller (due to a piston rod of large diameter) than the top area, their ratio going up to 10 : 1.

If the flow of the pump is going to the small area, driving the piston up with corresponding high velocity, the upper area expels a much larger flow, where the ratio of these flows is determined by the area ratio. This flow escapes over the directional control valve or over a non-return valve in parallel to avoid too high pressure losses.

Whilst the open circuit is mostly suitable for small presses, the *closed circuit* with reversible hydrostatic machines is preferable for large presses. The main piston has a vertical axis and descends during the working stroke, whilst on return only its weight has to be carried by the oil pressure. Fig 1 represents a closed circuit with the following notable features:-

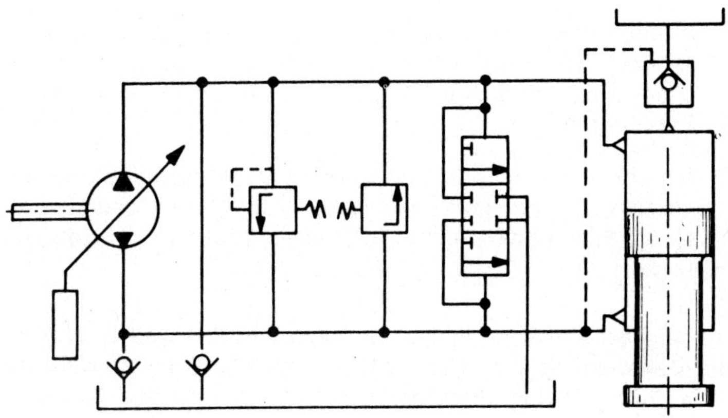

Fig 53.1: Hydraulic press with closed circuit and reversing hydrostatic machine

1.  Both main conduits are equipped with non-return valves and with connections to the tank, in order to avoid cavitation.

2.  There are two pressure relief valves in the main conduits for limiting the system pressure.

3.  The return piston area is much smaller than the forward area, with the large piston rod diameter, leading to a rapid return stroke. On the working stroke the small return area expels only a small flow, and the hydrostatic machine must be addit- ionally supplied by the non-return valve in the lower main conduit.

4.  The weight of the press piston is taken up by the pressure on the return piston face at the beginning of the descent. Its velocity is therefore determined by the return piston area and by the flow intake of the hydrostatic machine. The forward or upper piston area must therefore be supplied with a much larger flow over the big non-return valve, also called the prefill valve, from the small oil tank on top of the press cylinder.

This duty could theoretically be assumed by the non-return valve in the upper main conduit, but in practice the much larger cross-section of the prefill valve is required. With the start of the press action, the pressure in the upper cylinder space increases and the prefill valve closes.

5. A large flow is expelled by the upper piston area on the return stroke. The prefill valve is unlocked by the pressure in the connection, shown in broken lines, returning the excess flow into the tank.

6. Finally there is a decompression valve establishing a connection without pressure between the main conduit and the tank, thus preventing simultaneous pressure on both piston faces.

The highest pressure forces are experienced normally at the end of the downward stroke. The large volume above the forward piston face in this position therefore contains considerable energy with the compressibility of the oil. In order to initiate the return stroke, the hydrostatic machine is set for reverse flow. It takes the flow from the upper cylinder space, producing there a controlled pressure decrease. The hydrostatic machine works as motor and returns the torque to its drive, regenerating a large part of the compression energy. The main advantage of the closed circuit is that any sudden decompression of the upper cylinder space by the opening of a valve is avoided. This decompression would otherwise produce shock waves (treated in chapter 7).

Considering now the pressure in the lower main conduit during the beginning of the return stroke, it would increase very much quicker under the flow of the hydrostatic machine because of its much smaller volume. This is prevented by the decompression valve, which returns the flow into the tank. Only after the pressure in the upper main conduit has been reduced to almost zero, the decompression valve disconnects the lower main conduit from the tank, allowing pressure to build here and to initiate the upward movement of the press piston. Similar but less pronounced conditions apply at the end of the return stroke, allowing us to use a one-sided decompression valve in some simple installations.

Fig 2 represents a schematic cross-section of a decompression valve. The bores with the valve spool have the connections to the main conduits on top and the tank connections at the bottom. The cross bores, as represented, connect the end faces of the valve spool with both main conduits. It is pressed with pressure in the right main conduit to the left connecting the left main conduit with the tank and with pressure in the right main conduit it establishes a connection between the left main conduit and the tank.

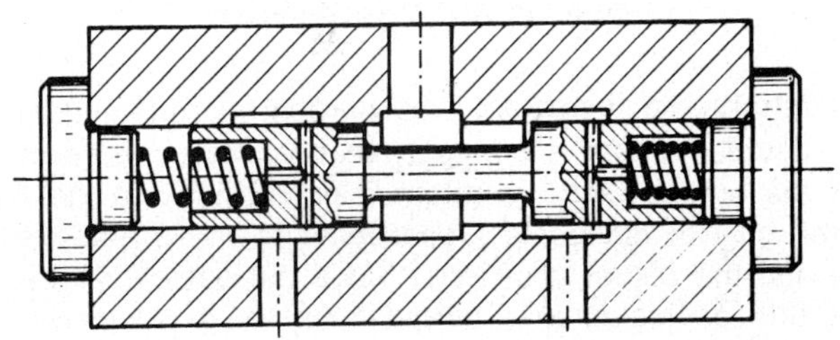

Fig 53.2: Schematic cross-section through a decompression valve

The decompression valve is usually made with closed centre, so that without pressure difference both main conduits are blocked. Sometimes small springs between valve spool and cover are used in order to locate it in the centre in the absence of pressure.

Both oil tanks shown in fig 1 are usually replaced in practice by a common tank. The common tank is frequently arranged on top of the press frame since it has to fill the upper cylinder space by gravity through the prefill valve. The top of the press frame also frequently contains the hydrostatic machine and the electric drive motor leading to a compact arrangement. Hydraulic presses also have many special problems, such as control circuits or shock waves or questions of optimum metal deformation, but they are outside the scope of this book.

## 54      Servocontrols

### Building and classification

Servocontrols an important application of oil hydraulics, are actuating devices where the output, that is the position of actuator piston or the output shaft, is returned or fed back to the input. The flow of the control device is proportional to the difference between the input and the output positions. Since the output closely follows the input this device is also called a follow-up servomechanism or servomotor, and its purpose is simply to produce an amplification of force or torque.

Servocontrols can be classified as follows:

1. Double acting servocontrols with valve and actuating cylinders.

2. Single acting servocontrols with valve and actuating cylinders.

3. Double acting servocontrols with variable displacement hydrostatic machines.

In principle it would be possible to design single acting servocontrols with hydrostatic machines using such machines with only one direction of pressure, but this has not found application in practice.

Valve controlled servocontrols are simpler in design, especially if they are single acting and a control valve has a more rapid response than the displacement of a hydrostatic machine. Their disadvantage is the large pressure drop in the servovalve with the consequent power loss. With hydrostatic machines, the power loss of servocontrols is comparatively small.

Fig 1 shows the basic design of a double acting servocontrol. We find at left the control or servovalve in the form of a spool valve, supplied by the connection in the centre left. The valve spool connects both sides of the actuating cylinder with supply or return depending on its position.

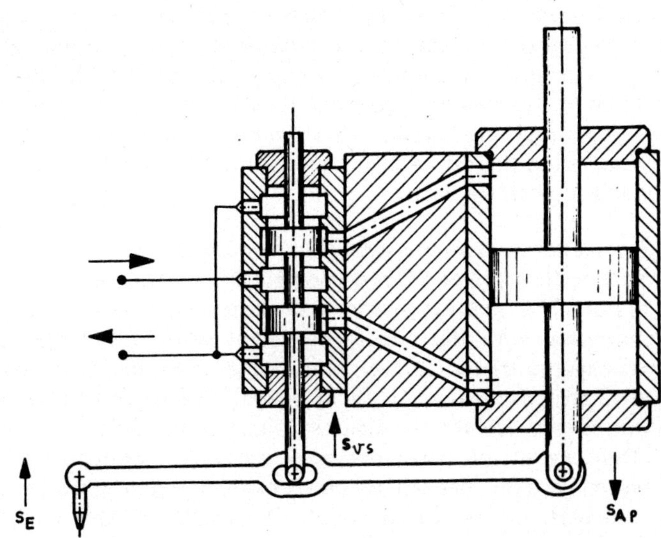



**Fig 54.1: Schematic of a double acting servocontrol
with feeler and horizontal feedback lever**

The horizontal lever connecting the rod of the actuating piston, the
valve spool and the input at left, also called a *feedback lever* is an
important component of this servocontrol. It is set up in such a way,
that any deviation of the valve spool from its central position admits
flow to the piston and imparts to it a motion, which causes the feedback lever to return the valve spool to its central position. This
arrangement is therefore a follow up mechanism, where the piston
runs opposite to the input point. We can also represent its operation
by saying that the valve spool remains in the steady state at its
central position and that the feedback lever follows the position of
the input with the valve spool as fulcrum.

The valve spool is similar in electrohydraulic servovalves with
the design of fig 45—3 and with the characteristics of fig 45—6. Both
cylinder spaces are generally under one half of the supply pressure,
if the servocontrol is at rest.

Any disturbing force on the piston rod has little return effect on
the input, since the piston is blocked by the two oil columns in the
actuating cylinder. In other words, the piston in the actuating cylinder is supported as on two oil columns. They will change their length
a little under disturbing forces due to the oil compressibility. As

mentioned in section 12, an oil column loses only 1% of its length with a pressure of 150 bar, if the cross-section is maintained constant. Consequently, if a force producing 150 bar attacks the actuating piston in its central position, it will displace only by 1% of half the length of the stroke. This figure is increased to 1.2% to 1.6% by including the oil volume under variable pressure between the cylinder and the control valve.

Fig 2 represents a single acting servocontrol, where the cylinder is equipped with a piston of two unequal areas, also called a differential piston. The smaller piston area around the piston rod is directly connected with the supply pressure, while the larger piston area is connected by the control valve either with supply pressure or with the return, depending on its position. The control valve of the single acting servocontrol is simpler, since it has only two metering edges. They are at the lower piston of the valve spool in fig 2. The upper piston is only needed to compensate the static force of the supply pressure on the valve spool. The small vertical connection with the return prevents a pressure build-up behind the upper piston of the valve spool. This design is also frequently called a half area servomotor since the upper piston area under constant supply pressure is usually one half of the lower area under variable pressure. There is always supply pressure in the space between the two pistons of the valve spool, while both outer faces of the valve spool are connected with the return, balancing its hydrostatic forces. In

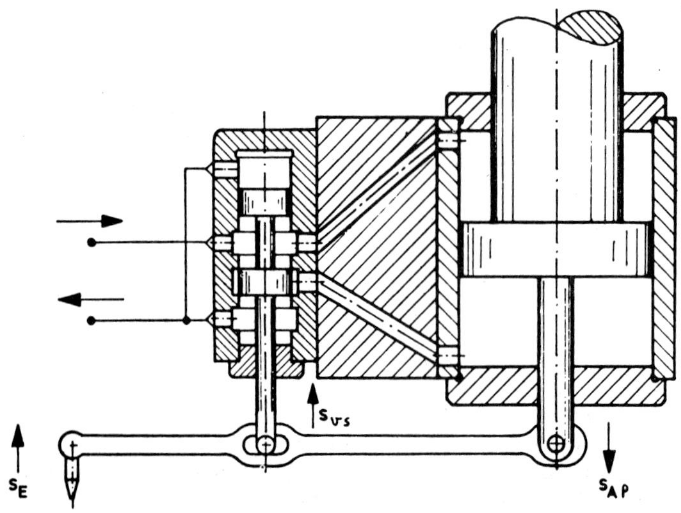



**Fig 54.2: Principle of a single acting servocontrol**

contrast, the hydrodynamic forces, or the forces due to the momentum carried by the oil jet, are uncompensated and are described in section 41.

Fig 3 shows a servocontrol with a hydrostatic machine with variable displacement and with an actuating cylinder as output. The feedback lever is vertical, whilst input and output are horizontal movements. The difference of these movements determines inclination of tilt of the centre of the feedback lever, changing the displacement setting of the hydrostatic machine.

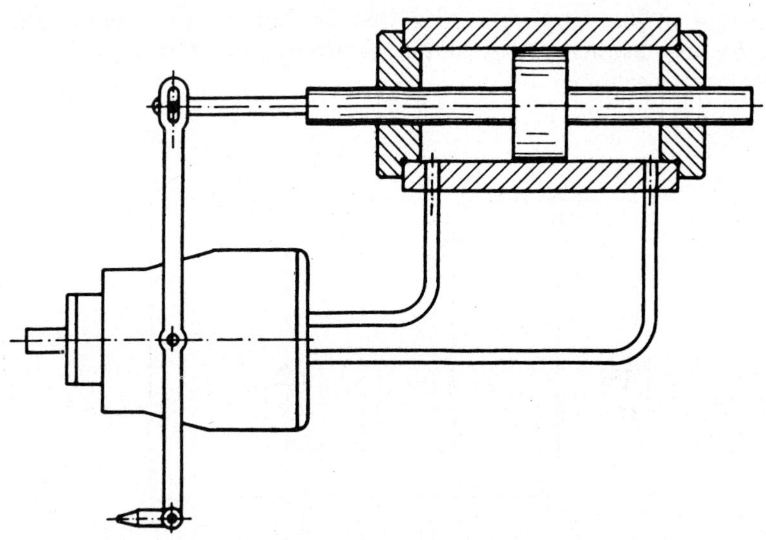

**Fig 54.3:** Principle of a servocontrol with adjustable displacement hydrostatic machine and with vertical feedback lever

## Operation of Servocontrols

We shall now study the operation of the servocontrol beginning with the functional diagram in fig 4 /FN'/. The position of the lever

FN[1] — Functional diagrams have been introduced for the study of such systems and are especially suited for it.

$s_E$ is the input variable and the difference $s_{sv}$ of the input and output position $s_{Ap}$ produces the flow Q by the first block and the velocity v of the actuating piston with the area A by the second block. The functional diagram contains an internal feedback loop with a first block giving the relation between velocity and force of the load. It carries an interrogation mark, since this relation is not very well known. The opposing force of the load produces the pressure by division with the area of the piston as indicated by the second block of the internal loop. This pressure reacts on the first block of the upper line, influencing the flow with given position of the valve spool. This first block therefore contains the relation between the position of the valve spool, the flow and the pressure, or in other words the characteristics of the servovalve or of a variable displacement machine.

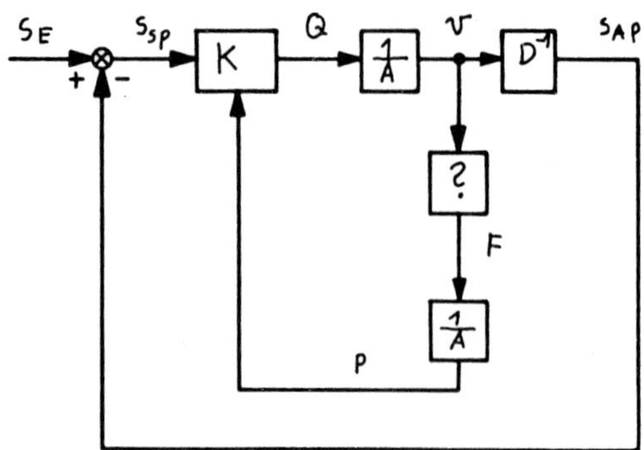


Fig 54.4: Basic functional diagram for a servocontrol

The external loop continues with a block with the operator $D^{-1}$, showing the change from the velocity into a position by integration. The position of the actuating piston is fed back to the subtraction point by the external connection. In reality, the subtraction takes place in the feedback lever. This basic functional diagram is valid for all kinds of servocontrols.

It is especially important to know the operation of the actuating piston with small variations from its steady state. The mass of the load constitutes a mechanical oscillator, together with the compliance or compressibility of the oil column in the actuating cylinder, and the consequent resonance possibility limits the speed of response of the servocontrol.

The functional diagram in fig 5 is valid for small deviations of the steady state, assuming that the inertia (or acceleration) forces of the load are preponderant. The first block gives the ideal flow of the control valve. The second block determines the velocity as before. Then begins the internal feedback loop, with the first block indicating the force needed by the actuating piston, ignoring any other but the inertia forces. This is obtained from the acceleration as derivation of the velocity, and then by multiplication with the mass of the load.

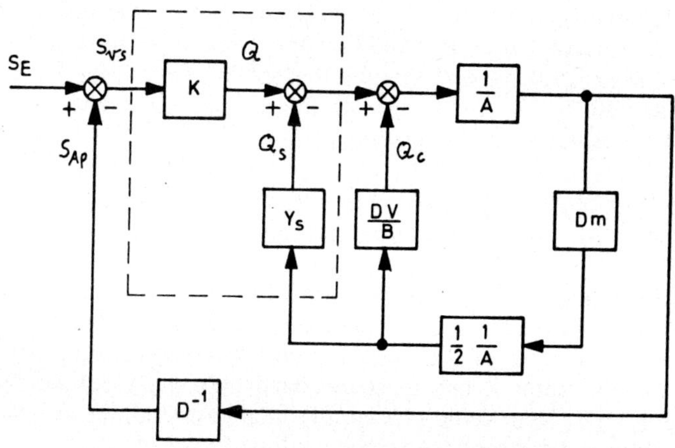


Fig 54.5: Functional diagram of a servocontrol for small deviations from the stationary state

In order to determine the relation between incremental force and pressure variation in one conduit, we must consider that such a variation is accompanied by an equal and opposite pressure variation in the other cylinder space, if the piston is in the central position.

Therefore each variation is only half as large as would correspond to the area of the piston, as expressed by the gain ½ A in the block below at right centre,/FN²./ The pressure variations produce the flow reduction $Q_s$ (similar to leakage in hydrostatic machines) of the

FN² — This is valid as long as the pressure variations are not so large as to produce cavitation in the low pressure cylinder space. In this case the stiffness would be reduced to one half for the exceeding part or the factor ½ above would have to be corrected or removed with consequent reduction of resonance frequency.

servovalve and the compressibility flow $Q_c$, by the two blocks in the centre, both deducted from the ideal flow.

It should be noted that $Y_s$ in fig 5 denotes the admittance of one pair of metering edges, metering in or metering out, and V the volume of one cylinder space with the piston in the centre position. To this the volume in one conduit to the control must be added. The volume in these conduits amounts to (after Guillon B 7 in section 733) about 50% to 60% in small and only 20% in large actuating cylinders. In consequence, the volume to be entered on the functional diagram fig 5 is 1.2 to 1.6 times as large as the volume of the area of the actuating piston and half the stroke. These remarks show that the actual volume and thereby the speed of response, is determined essentially by the length of the stroke, and that it is very difficult to reduce it by design changes.

The two blocks of fig 5 containing functions inside the control valve have been surrounded by the broken line for clarity.

Single acting servocontrols are described in principle by the same functional diagram of fig 5. If the symbols V and A denote now the volume and the area of the large piston (of the lower cylinder space in fig 2), the factor ½ has to be removed from the block below right centre of fig 5. This would change the loop gain and, as we shall see below, reduce the resonance frequency by a factor of $\sqrt{2}$

For a fair comparison of the resonance frequencies of single and double acting servocontrols, we must consider that for the same maximum force at given supply pressure the large area and the volume must be twice as large in the single acting case, since the cylinder space is simply piston area times length of stroke. In fig 5 we would, whilst removing the factor ½, have to replace A by 2 A and V by 2 V, leaving the loop gain unchanged. Therefore a single acting servocontrol of the same maximum force has twice the piston area but the same resonance frequency.

Fig 5 is also valid for servocontrols with variable displacement hydrostatic machines. As we shall see in chapter 6, in one conduit there will always be a constant low pressure set by the priming pump through the feeding valves, and therefore the factor 2 in the block below right centre of fig 5 must be removed. The blocks surrounded by the broken line then contain the characteristics of the hydrostatic machine, similar to that shown on fig 46–3. The tangent admittance represents, as we know, the inclination of these characteristics.

Instead of the pressure drop in the control valve, hydrostatic machines produce a loss pressure, as described in section 32. The consequent power loss is much smaller in servocontrols with hydrostatic machines. Using a hydrostatic motor for output gives the further important advantage that the volume under variable pressure can be made smaller than with an actuating cylinder. This is particularly true if both hydrostatic machines are arranged in proximity, in order to shorten the main conduits between them. In other words, the volume does not have the lower natural limit of one half of the length of the stroke and of the area of the actuating piston.

For completeness we shall apply control engineering methods for further study of fig 5, mainly by replacing the operator D by the variables of the Laplace transformation. This enables us to resolve the functional diagram, either by the operation described in appendix 2, or by the fomula of Mason. Determining first the transfer function of the inner loop

$$f_{in} = \frac{v}{s_{sv}} = \frac{K}{A} \; \frac{1}{1 + \dfrac{Y_s\, m}{2 A^2}\, s + \dfrac{m\, V}{2\, B A^2}\, s^2} \tag{54--1}$$

which can be written by introducing $\omega_n$ and $\zeta$ in the so called canonical form

$$f_{in} = \frac{K}{A} \; \frac{1}{1 + 2\,\zeta\, \dfrac{s}{\omega_n} + \dfrac{s^2}{\omega_n{}^2}} \tag{54--2}$$

The transfer function of the entire servocontrol becomes therefore

$$\frac{s_{Ap}}{s_E} = \frac{f_{in}/s}{1 + f_{in}/s} \tag{54--3}$$

with $\quad \omega_n{}^2 = \dfrac{2\, B A^2}{V m} \qquad\qquad \zeta = \dfrac{Y_s}{A}\,\sqrt{\dfrac{B m}{8 V}}$

The symbol $\omega_n$ in equation 2 represents the resonance frequency of the actuating piston on the oil columns. We see that the square of the resonance frequency is essentially the reciprocal loop gain of the compressibility term, except for the operator D. The resonance frequency limits the possible speed of response to disturbing forces. The symbol $\zeta$ denotes the damping, determined mainly by the admittance of the control valve. Naturally there are other damping effects, like the friction of the actuating piston. Nevertheless, the natural damping of hydraulic servocontrols is usually insufficient, leading to pronounced resonances. Therefore one sometimes uses artificial damping as described by Guillon, B 7. Finally it should be noted, that all these calculations cannot claim high accuracy, particularly since the admittances depend not only on the operating point, but also on the viscosity and on the temperature of the oil.

## Design and Dimensioning

The maximum required force determines, together with the admissible pressure, the piston area, which in turn imposes, together with the required velocity, the necessary flow. According to Guillon, B 7, aircraft control systems normally include a margin of only 10%, or the product of supply pressure and piston area is selected to 110% of the highest required force. For industrial servocontrols it is usual to include a higher margin up to about 50%.

Servocontrols with single acting control valves have the following limitations:

1. The larger piston area must be twice as large for a given force capacity at the same supply pressure, since it has to work in opposition to the smaller one, leading to more friction and more moving mass.

2. The metering edges and the cross-sections of the ducts must be larger with a given actuating velocity, due to the larger piston area.

The resonance frequency given by equation 2 is applicable with double acting servocontrols only as long as the actuating piston is in its central position. In other positions one volume increases and the other one decreases, leading to more stiffness of the oil columns and thereby increasing the resonance frequency, as treated more particularly by Guillon, B 7.

With single acting servocontrols, the position of the actuating piston affects the resonance frequency differently, and the above considerations are valid for the central position with one oil column under variable pressure. The resonance frequency increases, if this oil column becomes shorter, when the actuating piston in fig 2 moves downward. On the other hand, if the piston moves upward, the volume of the oil column becomes larger, up to twice as large at the upper end of the stroke. Since the servocontrol also has to work satisfactorily in this unfavourable position we can only count on a minimum resonance frequency of about 70% ( $= 1/\sqrt{2}$ ) of the resonance frequency of a double acting servocontrol with the same maximum force.

The single acting servocontrol is frequently manufactured because of its simplicity. Fig 6 shows as an example a hydrostatic machine with such a single acting servomotor as force amplifier with the

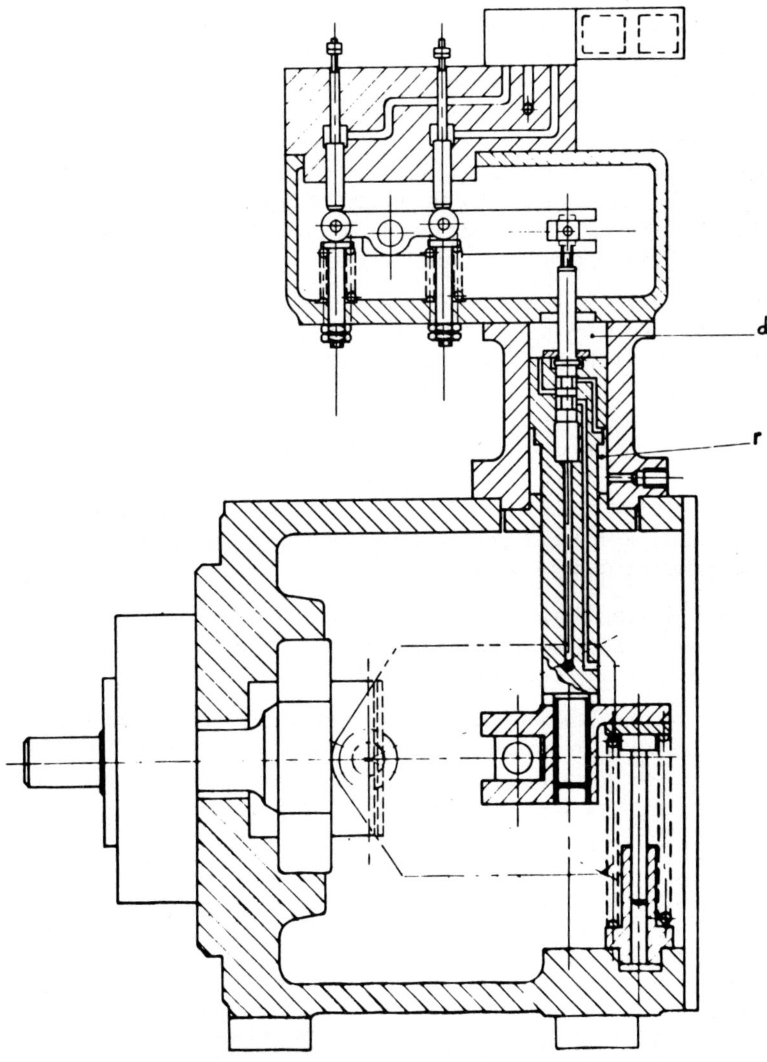



Fig 54.6:Axial piston machine with servocontrol for force amplification in displacement setting (Galdabini)

control valve arranged on the axis of the actuating piston for the displacement setting. This eliminates the feedback lever, since the opening of the metering edges is determined directly by the difference of the position of the valve spool and the actuating position. Generally this figure shows an axial piston machine in a casing below, with a displacement setting device in the upper bridge and with a simple acting servomotor in the connection between casing and bridge. The displacement setting in the bridge uses rockers that are brought into a fixed position by small pistons, fed with low or medium pressure over electric directional control valves. Other displacement setting parts can be used, but their force capacity is usually so small as to require a servomotor for force amplification. This example also shows that the connections or conduits in the simple acting servomotor are much simpler than with the double acting one.

Fig 7 represents a servomotor consisting of a hydrostatic axial piston machine with a control valve. The input shaft is on the left and the output shaft on the right. The admission and delivery of the working fluid to the cylinder of the axial piston machine is governed by the difference of angular position of both shafts, by means of the milled ports in the middle. Servocontrols including this servomotor are especially suitable for the feeds of numerical controlled machine tools, where the control signal on the input shaft is usually supplied by a stepping motor, that is a motor advancing through a certain angle on each electric control impulse.

Speed governors for water and other turbines are another important application of servocontrols or servomotors, where they are used as force amplifiers for actuating huge water valves.

Servocontrols are also very often used in production engineering, e.g. as copying devices on lathes or milling machines. A feeler travels on the template and moves the cutting steel over the force amplifier, in an arrangement similar to that represented in figs 1 and 2 above.

Servocontrols with hydrostatic machines are preferred in larger installations because of their reduced power consumption. The rudders of large ships are an important application, with actuation either by horizontal cylinders, or by a rotating vane on the vertical rudder shaft, similar to a vane pump. In very large ships ('Queen Mary', 1938) several servomotors are used in cascade; there the displacement of the largest hydrostatic machine is set by an actuating cylinder supplied by a smaller hydrostatic machine.

Hydrostatic motors with a rotary output shaft frequently replace the actuating cylinder. They constitute, really, a hydrostatic transmission built into a servomotor, as described more particularly in B 14.

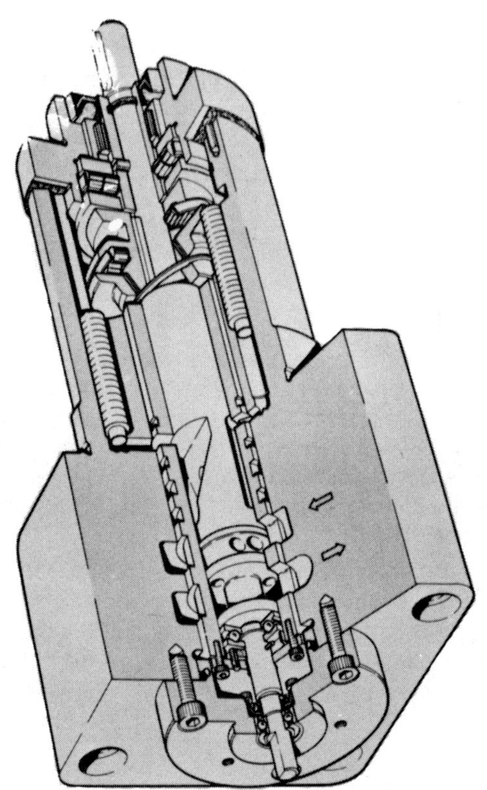

Fig 54.7:  Servocontrol with axial piston machine and installed mechanical servovalve (SIG)

# 6  HYDROSTATIC TRANSMISSIONS

## 61    Fundamentals and Circuits

Hydrostatic transmissions employ hydrostatic motors with rotary output, instead of actuating cylinders with linear output. They have found a large range of applications in vehicles, winches, machine tools and elsewhere. In spite of the differences due to the requirements of each application, there are a number of common problems that shall be discussed here.

Fig 1 represents a fundamental circuit of a hydrostatic transmission, with a variable displacement pump driven by a suitable prime mover with constant rotation frequency. It supplies oil to a similar hydrostatic motor on the left of fig 1. The rotation frequency and the torque of the pump are designated with $\omega_1$, $M_1$, and of the motor by $\omega_2$, $M_2$.

A hydrostatic transmission works like an electric transformer, as we shall see below. Therefore we shall call the hydrostatic machine coupled to the input shaft the 'primary', and the hydrostatic machine on the output shaft the 'secondary'. In normal operation, the primary works as pump and the secondary as motor, but these roles are interchanged if the output shaft is braked by the transmission.

The main flow between primary and secondary in fig 1 is proportional to the displacement setting of the primary.

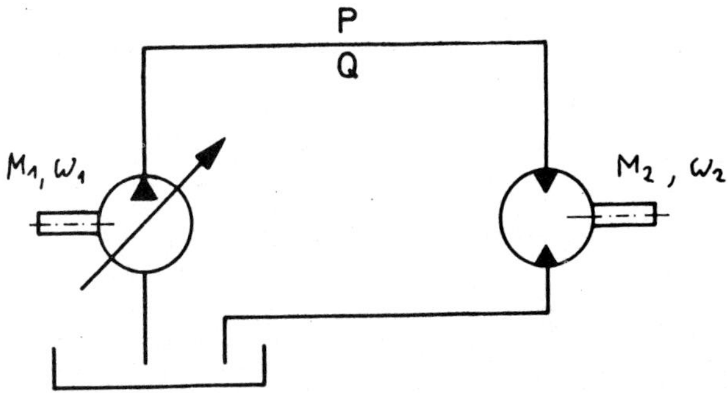


**Fig 61.1:** Open circuit hydrostatic transmission

The connection for the main flow between primary and secondary in fig 1 will be called the main conduit, since it transports the power. The oil returns from the secondary by the short connection into the tank. Due to the free surface, the tank makes sufficient oil available for the intake of the primary.

Hydrostatic transmissions with open circuit have the following limitations in their application:-

1.  They can produce output torque only in one direction, therefore they can neither brake the load nor drive it in reverse.

2.  The maximum rotation frequency of the primary is limited, due to the possibility of cavitation in the intake.

Due to this limitation, the control of the displacement setting of the primary must prevent the inversion of the direction of the output torque. This inversion is already produced with a load of large moments of inertia by quickly reducing the displacement setting of the primary and the output rotation frequency. The load would then overtake the selected output rotation frequency and the secondary takes more oil out of the main conduit than the primary supplies, leading to cavitation.

A closed circuit avoiding these limitations is built by replacing the open return through the tank in fig 2 by another main conduit, but some valves are necessary to ensure satisfactory operation of the closed circuit.

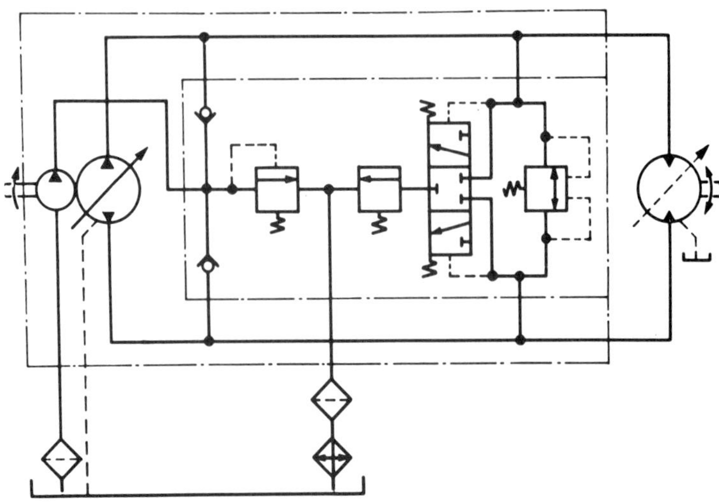

Fig 61.2: Closed circuit hydrostatic transmission

The circuit in fig 2 has the following notable features:

1. Each main conduit is equipped with one feeding valve, also
   called a makeup valve. These valves are simply non-return
   valves connecting either with the tank or preferably with the
   low pressure side of an auxiliary pump usually called a priming pump, driven by the shaft of the primary in fig 2. The
   priming pump takes oil from the tank over a filter to the feeding
   valves connected one to each main conduit. A relief valve
   limits the priming pressure to a value to 3 to 5 bar (50 to
   70 psi) and the flow of the priming pump is usually 10 to 15%
   of the maximum flow of the primary.

2. Fig 2 also contains main pressure relief valves set usually to
   250 to 350 bar (3 800 to 5 000 psi) on both main conduits. If the
   set pressure is exceeded, the relief valve opens a connection
   not to the tank, but into the other main conduit. This prevents
   cavitation if flow through the relief valve is higher than the
   flow of the priming pump.

3. An important detail is the rinsing valve, similar to the decompression valve in hydraulic presses (fig 53–2), shown in
   fig 2 in the form of a directional control valve. It opens a flow

path from the main conduit under low pressure, but a small relief valve downstream retains a certain minimum pressure. This relief valve therefore takes a certain flow from the main conduit under low pressure, and must be set somewhat lower than the priming pressure relief valve for correct operation.

4. The outflow of the rinsing valve is conducted over a filter and over a cooler to the tank, thereby rinsing the liquid.

As indicated by the broken line arrow in fig 2, a variable displacement secondary is sometimes used. The figure also has a point broken line around the valves which are arranged in a common valve block. The other point broken line in fig 2 comprises the primary with the priming pump and the valve block, outside of which there is only the secondary, both filters, the cooler and the oil tank.

Fig 3 shows a hydrostatic transmission, with a closed circuit. Primary and secondary have a displacement of 16 cm³/rd with a maximum frequency of 160 rd/sec. This corresponds to an installed power of 25 kW and to an apparent power, as defined in section 63, of 80 kW. The fixed displacement secondary is equipped with a flywheel for testing.

Fig 61.3: Hydrostatic transmission for 80 kW apparent power (Galdabini)

The quantitative operation of hydrostatic transmission without losses is described by the equations 31–1 and 31–2. Assuming for simplicity, that the maximum displacements of primary and secondary are equal, we obtain from the condition of equal flows of primary and secondary

$$Q = a\,\omega_1\,q_0 = \omega_2\,q_0 \qquad\qquad (61\text{–}1)$$

$a$ = displacement setting of the primary

Furthermore the pressures in primary and secondary are equal, /FN¹/.

This gives a relation for the torques

$$P = \frac{M_1}{a\,q_0} = \frac{M_2}{q_0} \qquad M_1 = a\,M_2 \qquad\qquad (61\text{–}2)$$

As already mentioned in section 31, the effective pressure in a closed circuit is equal to the difference of the absolute pressures between both main conduits. Therefore negative pressures are possible as well as negative torque and rotation frequencies. This corresponds to a braking of the output shaft, or to drive or braking in the reverse sense. For details consult Thoma, B 14 chapter 2.

The displacement setting $a$ in equations 1 and 2 can vary between + 1 through 0 to − 1. Consequently the output rotation frequency can change from full forward to stop to full reverse, reaching at most the input rotation frequency in both directions.

For a larger range of output frequencies it is necessary to use a secondary with adjustable displacement. For the quantitative treatment we use $a_1,\ a_2$ for the displacement settings of the primary and the secondary respectively and obtain as above from the equality of the flows

$$a_1\,\omega_1\,q_0 = a_2\,\omega_2\,q_0 \qquad\qquad \omega_2 = \frac{a_1}{a_2}\,\omega_1 \qquad\qquad (61\text{–}3)$$

---

FN¹ — If the pressure losses in the main conduit between primary and secondary can be neglected and from the equality of the pressures

$$\frac{M_1}{a_1 q_0} = \frac{M_2}{a_2 q_0} \qquad\qquad M_1 = \frac{a_1}{a_2} M_2 \qquad\qquad (61\text{--}4)$$

Hydrostatic transmissions work like electric transformers with an adjustable number of coil windings. They connect torque and rotation frequency of the input shaft with the same variables of the output shaft, as described by equation 1–4. It must be noted, however, that the actual values of the variables depend essentially on the properties or characteristics of the load and of the drive of the primary.

Apart from the simple circuit with one primary and one secondary, more complicated circuits with several primaries and several secondaries can be built. The most frequent form is two secondaries in parallel, where the pressures are equal, but the flows are added. Assuming equal maximum displacement of the 3 hydrostatic machines we obtain from the pressure and flow balances

$$\omega_2 + \omega_3 = \omega_1 \qquad\qquad M_2 = M_3 = \frac{M_1}{a} \qquad\qquad (61\text{--}5)$$

Here $\omega_3$, $M_3$ are the rotation frequency and the torque of the third hydrostatic machine.

Equation 5 shows that two secondaries in parallel have a differential effect similar to that between the drive half axles of automobiles. Therefore both secondaries supply the same torque to their output shaft under their equal pressures. This differential effect is sometimes undesirable, especially with more than two secondaries in parallel. As an example, if one wheel of a vehicle starts to slip, its adhesion to the ground diminishes. Due to the torque equalisation, the drive effect of the other wheels is equally reduced. The sliding wheel assumes a high speed absorbing the entire flow, as is known from automobiles on ice. Flow control valves or flow dividers can be used to avoid large speed difference of secondaries in parallel, but they produce a pressure drop and losses. Furthermore the setting of the flow control valve has to be changed when changing the output rotation frequency.

Circuits with two or more secondaries in series have different properties. The flow and the output rotation frequencies are equal (neglecting the influence of leakage) as determined by the flow of the primary. Each wheel (or generally, load) requires a certain torque, and this determines the pressure drop on the corresponding secondary. The sum of these pressure drops gives the operating pressure of the primary. The series circuit therefore is quite satisfactory except for the disadvantage that the pressure on the primary becomes very high if torque is required simultaneously by both wheels. In addition certain hydrostatic machines are highly stressed by oil pressure in both main conduits. Therefore this is rarely used in practice.

Connecting several secondaries over a directional control valve with an open passage in the central position gives much more favourable stressings, especially if the full torque is normally required only of one secondary, as mentioned in section 44. This leads to very interesting combinations, including actuating cylinders. The pressure relief valves protecting the primary must not be omitted.

## 62    Operation and Characteristics

We shall now describe the operation of hydrostatic transmissions with functional diagrams. For simplicity, only the case of an adjustable primary and a fixed secondary of the same maximum displacement will be treated; for more complicated cases, the reader is referred to B 14.

The functional diagram in fig 1 shows the principle of operation of a hydrostatic transmission. The displacement setting $\alpha$ of the primary determines the flow $Q_i$ by means of the first and the output rotation frequency $\omega_2$ by the second block. Both blocks contain the fundamental equations 31—1 and 31—2.

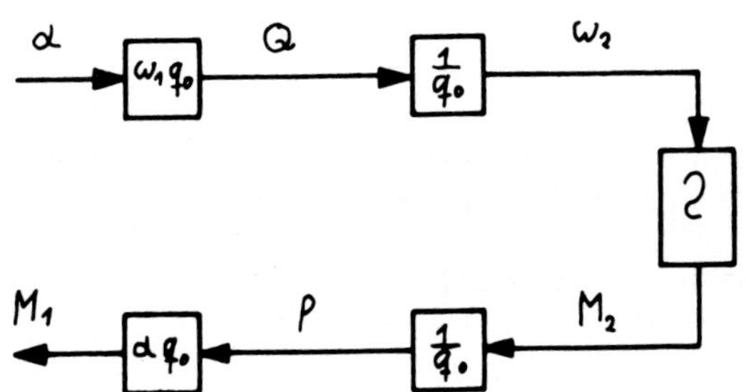

Fig 62.1: Basic functional diagram for a hydrostatic transmission

The relation between rotation frequency and torque of the load is usually rather indeterminate, as represented by the vertical block with the interrogation mark. The torque produces the operating pressure and the torque needed by the primary through the lower blocks. With this functional diagram, especially suitable for constant input rotation frequency, we have chosen a causality, i.e. the chain of causes and effects in a hydrostatic transmission.

For further treatment, we use the more accurate functional diagram in fig 2, giving the losses. The loss torque $M_{2L}$ of the secondary in the block below at the right influences the operating pressure.

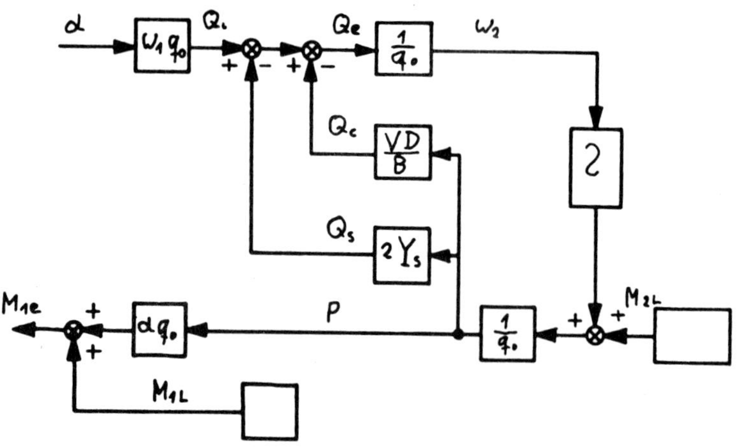

Fig 62.2: Functional diagram for a hydrostatic transmission with losses

This pressure determines the ideal torque of the primary to which the
loss torque $M_{1L}$ must be added in order to obtain the effective torque
that the primary draws from the prime mover.

The pressure also produces the leakage flow $Q_s$ in the central
block with the admittance $Y_s$ that is deducted from the ideal flow
$Q_i$, similar to that already described in section 54 with servomotors.
The factor 2 in this block indicates that the leakages of the primary
and the secondary are added.

Fig 2 also contains a block in the centre determining the compressibility flow $Q_c$. Here, V denotes the volume in one main conduit
and one pressure side of the hydrostatic machine, and B the compressibility modulus. The letter D denotes the derivation operator
indicating that compressibility flow and pressure are related as
follows

$$Q_c = \frac{V}{B}\frac{dp}{dt} = \frac{V}{B}D\,p \qquad\qquad (62\text{--}1)$$

The compressibility flow is thus proportional to the change of
pressure and becomes significant only with rapid pressure variations.

The functional diagram in fig 2 adequately represents the operation of a hydrostatic machine, since leakage and compressibility flow are small during normal operation, and therefore only produce a small correction to the ideal flow. In other words, the gain of the feedback loop, the loop gain in fig 2, is small compared to one.

For a more detailed treatment, it is necessary to know the characteristics of the load. If the output shaft is coupled to a large moment of inertia, with rapid variation of the output rotation frequency, the acceleration torque becomes preponderant. This torque can be written into the block representing the characteristics of the load, neglecting the other output torques. Thereby, a derivation of the output rotation frequency with respect to time appears, as in fig 54—5.

The literature contains other types of functional diagrams obtained by inversion of the feedback loop in fig 2. In this case, the compressibility flow is the variable responsible for pressure and torque. The derivations with respect to the time are then replaced by integrations and the loop gain of the feedback loop is very high, especially at low frequencies or slow variations of the displacement setting of the primary. While such an inverse functional diagram is correct in principle, it corresponds much less to a natural insight into the operation of a hydrostatic transmission and is, therefore, not normally used by the author.

All loss torques are functions of the different variables, as could be indicated by cross connections, omitted in fig 2 for clarity.

In order to determine the influences of the losses on the output characteristics of hydrostatic machines, we use the slip frequency and the loss pressure, as defined by our equations 31—5 and 31—6. These variables have the advantage that they are practically independent of the maximum displacement of the hydrostatic machine and are, therefore, almost universally applicable. In principle, they could be entered into a functional diagram similar to fig 2, by providing a few more blocks representing multiplication or division with the maximum displacement.

We shall use equation 31—7 with the subscripts 1 and 2 denoting the variables of the primary and the secondary respectively. From the balance of the flows, we obtain

$$\omega_1 \, a \, q_0 - \omega_{S1} \, q_0 = \omega_2 \, q_0 + \omega_{S2} \, q_0$$

which gives for the output rotation frequency

$$\omega_2 = a\,\omega_1 - \omega_{S_1} - \omega_{S_2} \tag{62-2}$$

with $\omega_{S_1}$, $\omega_{S_2}$ = slip frequencies of the primary and secondary defined by equation 31—5.

We see from equation 2 that the slip frequencies of the primary and secondary are simply deducted from the ideal rotation frequency of the output shaft.

We obtain a relation between the torques of the primary and secondary from the equality of pressure by using equation 31—8

$$\frac{M_{1e} - p_{L_1}\,q_0}{a\,q_0} = \frac{M_{2e} + p_{L_2}\,q_0}{q_0} \tag{62-3}$$

which gives by solving for $M_2$

$$M_{2e} = \frac{M_{1e}}{a} - \frac{p_{L_1}\,q_0}{a} - p_{L_2}\,q_0$$

where $p_{L_1}$, $p_{L_2}$ = loss pressures of the primary and secondary defined by equation 31—8.

The loss pressures normally will be different, since primary and secondary work at different rotation frequencies and displacement settings, as distinguished by the subscripts.

Equations 2 and 4 are very suitable to determine the effective performance of a hydrostatic transmission including the inevitable losses. Many manufacturers prefer to use the efficiency defined as ratio of output to input power. It has the advantage that it gives some indication for superficial studies but it is much less suitable to ascertain the performance of a transmission over a large range of variables. For completeness, we shall now deal with the relation between loss pressure and slip frequency with the volumetric and mechanical efficiencies $\eta_V$ and $\eta_m$ of a hydrostatic transmission. These efficiencies are defined as follows

$$\omega_2 = a\,\omega_1\,\eta_V \qquad\qquad M_{2e} = \frac{M_{1e}}{a}\,\eta_m \tag{62-5}$$

Comparing equations 2 and 4 gives

$$\eta_v = 1 - \frac{\omega_{s1} + \omega_{s2}}{\alpha\,\omega_1} \qquad \eta_m = 1 - \frac{p_{L1}\,q_0}{M_{1e}} - \frac{p_{L2}\,q_0}{M_{1e}/\alpha} \qquad (62\text{--}6)$$

Since the denominators in equation 6 are rapidly variable, the behaviour of the efficiencies over a range of operating variables is more complicated than the behaviour of the slip frequency and loss pressure, as described in section 31.

In experiments with hydrostatic transmissions, the ideal output frequency must first be determined. This is done best by running the transmission without output torque since then the leakage is presumed negligible. The reduction of the output frequency with increasing torque gives the volumetric efficiency, and the comparison of the primary and secondary torques, the mechanical efficiency, by equation 5.

## 63    Dimensioning of Hydrostatic Transmissions

In the preceding section, we have treated the characteristics, i.e. the relation between torque and rotation frequency of hydrostatic transmissions. As mentioned before, such characteristics must be clearly distinguished from the operating limits or the maximum admissible values of the different variables like rotation frequency and torque. These maximum values govern the dimensioning or the selection of a hydrostatic transmission for a given application.

We have seen in section 36 that there are two kinds of operating limits:

1. Intrinsic limits that are never exceeded by the machine.

2. External limits which can be exceeded by the machine resulting in danger.

Examples for the intrinsic limits are the maximum torque of an internal combustion engine, and the maximum output rotation frequencies of a hydrostatic transmission with variable primary, according to equation 61—1. On the other hand, the rotation frequency of an adjustable secondary described by equation 61—3 can become very high. It is thus an external limit which must be observed by the displacement setting devices, in order to avoid dangerous overspeeds.

A further external limit of a hydrostatic transmission is the possible torque (absorbed by the load) due to the admissible operating pressure. Here, the displacement setting of the primary must select the output frequency in such a way that excessive load torque is avoided. In general terms, the control of hydrostatic transmissions (or of all other machines) must provide that no variable exceeds its limits. In order to avoid damage in exceptional cases or with malfunctioning of the control, suitable safety devices, especially pressure relief valves, are used.

The dimensioning of hydrostatic transmissions is based on the graph of torque against rotation frequency of the load, as shown in fig 1. It also contains the characteristics of the secondary for four different displacement settings of the primary, represented by the continuous, almost vertical, lines. The slight inclination of these characteristics is due to the leakage in the primary and secondary.

Fig 1 contains the following operating limits important for the dimensioning of hydrostatic transmissions:

1. The maximum output torque determines, together with the admissible pressure, the displacement of the secondary.

2. The maximum rotation frequency of the load, the broken vertical line at the right, is an intrinsic operating limit.

3. Sometimes the admissible torque of the primary is also limited, not by the hydrostatic transmission, but by the prime mover. This produces the point-broken curve in fig 1, by equation 61–2.

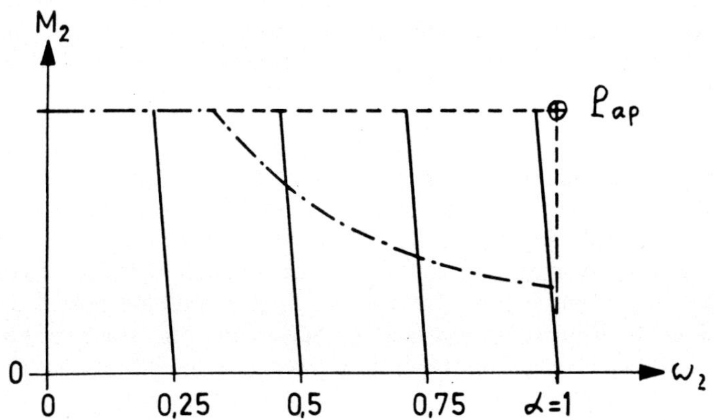


**Fig 63.1:** Characteristics of a hydrostatic transmission with the point representing the apparent power

In many cases, the maximum frequency of the secondary and of the load are different, requiring the installation of a fixed reducing gear.

It should be noted that the torque limit of the drive is independent of the hydrostatic transmission, which could also use the field above the point-broken limiting curve in fig 1. Some exception to this statement is to be found in the fact that very high operation pressures are admitted for some hydrostatic machines only during a part of the operating cycle. They should, therefore, be used only with small output frequency, i.e. with small displacement setting of the primary.

Furthermore, some hydrostatic machines support higher pressures with small displacement setting, especially swash plate axial piston machines, due to the radial forces described in section 33–5.

The torque limit of the prime mover is also often, somewhat improperly, called the power limit.

The dimensioning of hydrostatic transmissions is determined much less by the power of the prime mover than by the maximum rotation frequency and the torque of the output. These two variables can be combined in a variable of the dimension of a power. This combination, called *apparent power*, is obtained by multiplication of the maximum values of output torque and rotation frequency

$$P_{ap} = \omega_{max}\, M_{max} \qquad\qquad (63–1)$$

Equation 1 is similar to equation 31–4, since the apparent power of a hydrostatic transmission is determined by the apparent power of the secondary. It is, furthermore, independent of any reduction gear between secondary and load.

Fig 2 represents the torque versus rotation frequency of an automobile transmission, as an example of the possible performance of hydrostatic transmissions. The transmission can produce the upper curve with full engine rotation frequency and accelerator position if the transmission losses are neglected. Including these losses, we obtain the lower curve. The oblique arrows indicate how the points between both curves are displaced due to the slip frequency and the loss pressure. Fig 2 represents furthermore the maximum torque of a mechanical four-speed gear box in the same engine. It pertains to a small automobile with an admissible mass of 1 275 kg, a maximum speed of 140 km/h and an engine of 6.6 daNm and 600 rd/sec rotation frequency.

The required apparent power determines the displacement volume of the hydrostatic machine as well as its mass, bulk and cost. The loss torques, the leakage flow and the connected power losses increase proportionally with the maximum displacement volume, since loss pressure and slip frequency are almost independent thereof. If such a hydrostatic machine is driven by a weak prime mover, the relative power loss is large or the efficiency unfavourable. This can be seen from equation 62–6 since the denominator becomes small with small input torque. With a given apparent power of the load, it is therefore recommended not to select too weak a prime mover. The ratio between apparent power of the load and power of the prime mover

$$\gamma \;=\; \frac{P_{ap}}{P_1} \qquad\qquad (63\text{--}2)$$

is called the *torque multiplication*. For the reasons given, it is usually not economical to exceed values of 3–4.

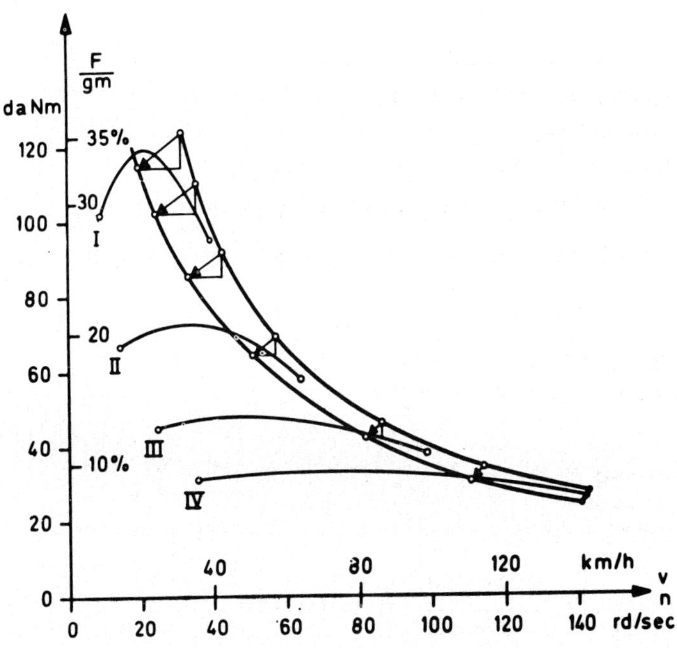


Fig 63.2: Characteristics of a hydrostatic transmission
for automobiles including the effect of losses and the
performance of a conventional 4-speed gear box

One of the difficulties in the application of hydrostatic transmissions is that many loads require torque multiplications in the range of 10–12. As an example, lorries and buses demand an apparent power of 1 600–1 900 kW for an installed power of 160 kW. Such a ratio demands not only large hydrostatic machines but also produces considerable power losses and unfavourable efficiency. In order to avoid these difficulties, it is sometimes economical to provide a mechanical two-speed gear box between the secondary and the wheels, as described more particularly in B 14.

### Numerical Example of the Dimensioning of Hydrostatic Transmissions

An application for a transmission with adjustable primary and fixed secondary has the following data

$$\omega_{2\,max} = 140 \text{ rd/sec} \qquad\qquad M_{2\,max} = 120 \text{ daNm}$$

$$\omega_1 = 420 \text{ rd/sec}$$

Neglecting the loss, we want to determine:

1. Displacement of the secondary

2. Maximum flow

3. Maximum displacement of the primary

4. Input torque $M_{1\,max}$ required to produce 100 bar with $\alpha = 1$.

The displacement of the secondary becomes by equation 31—2b

$$\frac{q_{20}}{1 \text{ cm}^3/\text{rd}} = \frac{M/1 \text{ daNm}}{p/100 \text{ bar}} = \frac{120}{3} = 40 \qquad q_{20} = 40 \text{ cm}^3/\text{rd}$$

and the flow by equation 31—1 with $\alpha = 1$

$$Q = \omega_2\, q_{20} = 140 \frac{\text{rd}}{\text{sec}}\; 40 \frac{\text{cm}^3}{\text{rd}} = 5600 \text{ cm}^3/\text{sec} = 5.6 \text{ lit/sec}$$

The maximum displacement of the primary according to equation 31—1

$$q_{10} = Q/\omega_1 = \frac{5600 \text{ cm}^3/\text{sec}}{420 \text{ rd/sec}} = 13.3 \text{ cm}^3/\text{sec}$$

For an operating pressure of 100 bar, we need after equation 31—2b

$$\frac{M_1}{1 \text{ daNm}} = \frac{p}{100 \text{ bar}}\; \frac{q_{10}}{1 \text{ cm}^3/\text{rd}} = 13.3 \; ; \; M_1 = 13.3 \text{ daNm}$$

The apparent output power is by equation 1

$$P_{ap} = 140 \text{ rd/sec} \cdot 120 \text{ daNm} = 16800 \frac{\text{daNm}}{\text{sec}} = 168 \text{ kW}$$

and the input power

$$P_1 = \omega_1\, M_1 = 420 \frac{\text{rd}}{\text{sec}}\; 13.3 \text{ daNm} = 5600 \frac{\text{daNm}}{\text{sec}} = 56 \text{ kW}$$

The torque multiplication becomes by equation 2

$$\gamma = 168 \text{ kW}/56 \text{ kW} = 3$$

quite an acceptable value for hydrostatic transmissions.

Let us now calculate the torque that can be delivered at maximum output frequency with $\alpha = 1$

$$\frac{M_2}{1 \text{ daNm}} = 1 \cdot 40 \qquad M_2 = 40 \text{ daNm}$$

**Influences of Transmission Losses on Dimensioning**

We shall now indicate how the losses influence the dimensioning of this example transmission. With small $\omega_2$ and 300 bar operating pressure, we can expect a loss pressure of 10 bar. Thus, 290 bar remain for the output torque giving a displacement of the secondary of

$$\frac{q_{20}}{1 \text{ cm}^3/\text{rd}} = \frac{120}{2.9} \qquad q_{20} = 41.4 \text{ cm}^3/\text{rd}$$

and an ideal flow of

$$Q_i = \omega_2 \, q_{20} = 140 \, \frac{\text{rd}}{\text{sec}} \cdot 41.4 \, \frac{\text{cm}^3}{\text{rd}} = 5\,800 \text{ cm}^3/\text{sec}$$

$$q_{10} = Q_i/\omega_1 = \frac{5\,800 \text{ cm}^3/\text{sec}}{420 \text{ rd/sec}} = 13.8 \, \frac{\text{cm}^3}{\text{rd}}$$

The slip frequencies must be considered for the determination of the primary, so that the required output frequency is obtained with $\alpha = 1$ and 100 bar operating pressure. The slip frequency of the secondary can be taken as 1.5 rd/sec at this pressure. The necessary effective flow of the secondary becomes according to equation 31—7

$$Q_e = (\omega_2 + \omega_{S2}) \, q_{20} = (140 + 0.5)\frac{\text{rd}}{\text{sec}} \, 41.4 \, \frac{\text{cm}^3}{\text{rd}} = 5\,870 \text{ cm}^3/\text{sec}$$

The slip frequency of the primary $\omega_{S1}$ should be about 2.5 rd/sec because of the higher rotation frequency, and we determine with equation 31—7

$$q_{10} = \frac{Q_e}{\omega_1 - \omega_{S1}} = \frac{5\,870 \text{ cm}^3/\text{sec}}{417.5 \text{ rd/sec}} = 14 \text{ cm}^3/\text{rd}$$

Summarising we note that the losses have only a very small influence on the dimensioning of hydrostatic transmissions. They can be neglected without hesitation for a rapid approximate determination of suitable transmission sizes for a given application.

Since calculations of this kind frequently occur in the application of hydrostatic transmissions, it is convenient to prepare a dimensional diagram. This is a kind of computational flow diagram, a little like a functional diagram, with blocks and connections showing the dimensioning relations treated in this numerical example. The numerical values or even several choices thereof can be written in prepared spaces beside the connections. Together with graphs of loss pressures and slip frequencies, it allows a very economic and accurate determination of hydrostatic transmissions.

### 64    Power Split Hydrostatic Transmissions

Power split hydrostatic transmissions are built by the combination of a hydrostatic transmission and a mechanical differential gear, similar to the rear axle of an automobile. They are often recommended because a part of the power is transmitted mechanically and a part hydraulically, reducing the losses. For a proper study, it is much better to consider the relation between torques and rotation frequencies, instead of the powers, in order to avoid unnecessary complications./FN[1]/We shall see that power split transmissions are especially favourable if the range of output rotation frequency is small.

Fig 1 contains the basic arrangement of a power split drive with one half axle of the differential driven by the input to which the primary is also coupled. The other half axle is coupled to the secondary and the output shaft of the differential drives the load.

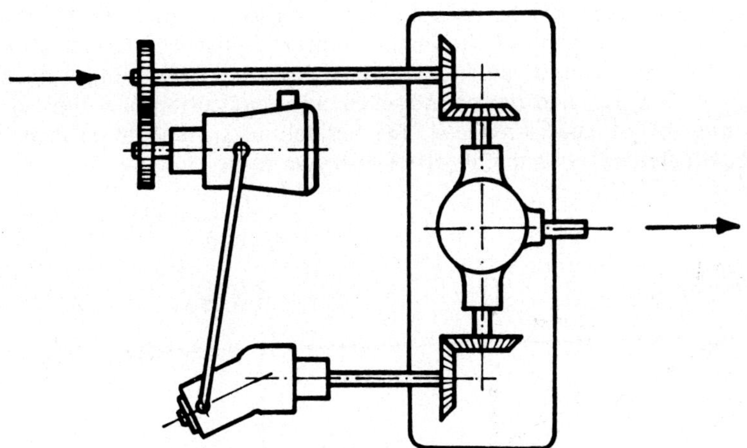

**Fig.64.1:**Principle of a power split transmission represented by an automobile differential

---

FN[1] — There exist many publications dealing with the very complicated power relations in such transmissions whilst the relations between torques and rotation frequencies, as developed here, are quite simple.

Denoting by $\omega_u$, $M_u$ the rotation frequency and the torque of the load, we have

$$\omega_u = \frac{\omega_1 + \omega_2}{2} \qquad\qquad M_u = 2\,M_2 \qquad\qquad (64\text{--}1)$$

Equation 1 describes the operation of a differential where the torques in the half axles are equal and the rotation frequencies are added. We immediately see the advantage that the output torque is twice as much as the torque of the secondary, allowing us to use a smaller secondary with a given load. Inserting the relation $\omega_2 = a\,\omega_1$ from section 61 into equation 1, we obtain

$$\omega_u = (a + 1)\,\omega_1 \qquad\qquad (64\text{--}2)$$

Here, $a$ can vary from $-1$ through 0 to $+1$. The highest output rotation frequency is equal to the input rotation frequency and obtained with $a = 1$, whilst $a = 0$ produces half the output rotation frequency. With $a = -1$, the output shaft remains stationary.

The advantage of the greater output torque is paid for by the reduction of the range of the output rotation frequency. These relations are represented graphically in fig 2 with the straight lines $\omega_{2min}$ and $\omega_{2max}$ and the point-broken lines indicating how the output frequency range is reduced. The variable $\omega_{2min}$ should be understood algebraically as the negative maximum value of $\omega_2$.

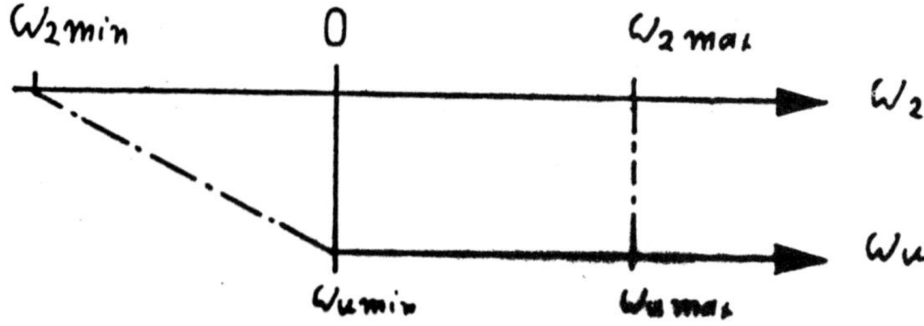


**Fig.64.2:** Limitation of output frequency change in a power split transmission

As we have seen, the power split transmission of fig 1 cannot reverse the rotation of the output shaft./FN²/ The apparent power is doubled compared to the simple transmission due to the increased output torque

$$\omega_{u_{max}} \; M_{u_{max}} = 2 \, \omega_{2max} \; M_{2max}$$

Coming now to the losses, the torque losses of the power split drive are caused by the torque losses of the hydrostatic machines influencing the output torque only through a factor of ½ by equation 1 and by the torque losses of the differential. In addition, the slip frequencies of the hydrostatic transmission must be simply deducted from the secondary frequency in equation 1, while the gears of the differential have no slip. The total losses are, therefore, smaller than in a comparable transmission with twice the hydrostatic apparent power.

Starting from the basic arrangement of power split in fig 1, all of the many variants can be obtained by using other reduction ratios in the gears shown in fig 1, as described in B 14. It should only be mentioned that the gain of the output torque increases with a smaller required range of output frequency. Power split transmissions are, therefore, especially favourable if only a small range of output frequencies are needed.

The above considerations about power split transmissions are based on their kinematic relations only and are therefore quite independent of constructional details. In practice, such transmissions are built from differentials, either with bevel gears or with planetary spur gears. In a variant, the gears are eliminated by allowing the casing of the hydrostatic transmission itself to rotate. Furthermore, the power split is not limited to hydrostatic machines but the differential can be driven by any other variable ratio transmission, such as mechanical friction drives. In the latter case, the differential sometimes serves the inverse purpose, namely to enlarge the range of output frequencies, resulting in a torque reduction instead of a torque gain.

---

FN² — This can be avoided by using a hydrostatic transmission with variable secondary, leading to more unfavourable torque conditions.

# 7  VIBRATIONS AND NOISES

## 71    Oscillations and Resonances

So far, we have assumed that the operation of hydrostatic machines proceeds smoothly with the desired pressures and forces. Many parts in oil hydraulics have not only a mass but are also somewhat elastic and can, therefore, be excited to oscillations.

We have already encountered, in the servomotors of section 54, resonances between the mass of the load and its elastic suspension. The elasticity came from the compressibility of the oil in the conduits and in the actuating cylinder. Such a system is frequently insufficiently damped and can oscillate strongly with only a weak excitation.

Other frequent disturbances are shock waves, especially pressure shock waves running through all components and hydraulic conduits, as we shall deal with more particularly in section 72. As the damping of these shock waves is frequently insufficient their presence in conduits can lead to substantial sound radiation.

Shock waves in hydraulic conduits can be produced by the following causes:

1. Shocks due to rapid change of the impedance of components, as on closing of valves.

2. Instability of valves as, for instance, oscillating pressure relief valves.

3. Periodic pressure variations in hydrostatic machines.

Here we should make the following comments:

1.  The shocks produced by changes of impedances can be reduced by slower speed, especially by slower closing of valves. Furthermore, such shocks are largely absorbed by hydraulic accumulators which are especially effective if they are installed at the point of maximum pressure shock.

2.  Valves with very weak springs are especially liable to instability. It is, therefore, necessary to use piloted pressure relief valves if characteristics with high tangent admittances are required. Furthermore, any change of hydraulic forces, such as the change of the jet angle with the position of the valve spool treated in section 41, gives disturbing forces. Finally, servovalves with very soft actuators of piloting devices and with high supply pressures are also subject to instability, as dealt with in B 1.

3.  The periodic pressure variations in hydrostatic machines are produced during the switch of the displacement chamber between the intake and delivery ports, i.e between low and high pressure. Figure 1 represents the pressure in a displacement chamber or cylinder as a function of time or of the rotation angle of the shaft. With instantaneous switching, the graph of the pressure in the cylinder has the form of a rectangle, as is shown in thin lines. There are very steep increases, producing shock waves in the entire installation. If the pressure in the cylinder increases more gradually, giving a trapezoid-like curve — in fig1 in heavy lines — the shock waves are much weaker. A condition for this gradual pressure increase is a

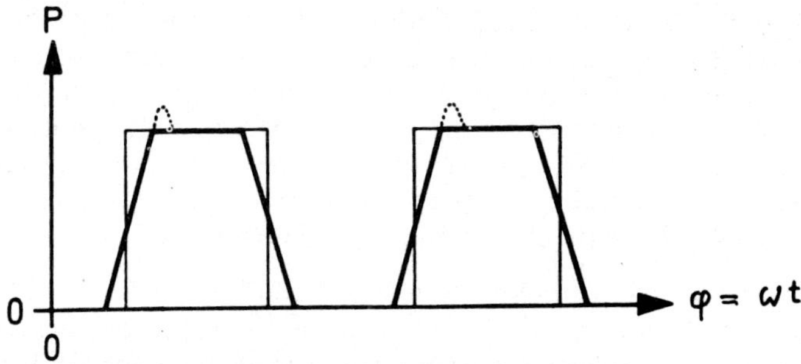


Fig 71.1: Pressure as function of time in the cylinder of
a hydrostatic machine

comparatively slow switching of the cylinder from low pressure to high pressure obtained by a large lap angle, as described in section 33—4. However, the lap angle is limited by the piston movement and the consequent influence on the pressure development must be controlled by suitable throttling grooves.

The steepness of the pressure rise is, therefore, essentially responsible for the excitation of oscillations in hydrostatic installations. We shall, therefore, quickly calculate the available time for the pressure rise in hydrostatic machines. The lap angle with a maximum of 10 deg = 0.17 rd gives an available time of

$$\tau = \frac{\phi}{\omega} = \frac{0.17 \text{ rd/sec}}{160 \text{ rd/sec}} = 1.06 \cdot 10^{-3} \text{ sec} \approx 1.1 \text{ m sec}$$

This shows that the noise production of hydrostatic machines increases with pressure and rotation frequency. Especially unfavourable are pressure peaks, as shown in pointed lines on fig 1. They are produced by incorrect adjustment of the distribution or by too large lap angles with unfavourable throttling grooves.

Closely related, are measures against the noise of hydrostatic machines and installations, which can be classified as follows:

1. At the source, by improvement of the working diagram or pressure graph.

2. By preventing noise radiation with elastic suspension of vibrating parts.

Whilst it is always preferable to reduce vibrations at the source, as described above, the elastic suspension of hydrostatic machines also brings worthwhile gains by preventing vibrations from being transmitted to the casing and radiated as sound. It is important to reduce sound reflections within the casing as much as possible. Furthermore, by using stiff walls and not too large plates for the casing, astonishingly good results can be obtained.

Conduits and hydraulic connections can be a problem with elastic suspension of hydrostatic machines. For the necessary deformations, one uses either rubber hoses or several steel tubes of small diameter in parallel, in order to allow easier bending. Similar procedures are known from electrotechnics where cables usually consist of many strands in parallel for flexibility.

### 72      Shock Waves on Valve Closure

We shall now consider a flow through a hydraulic conduit of constant cross section with the mean velocity v and the tube terminated
by a valve at the right, as shown in fig 1. Supposing that the valve
closes suddenly at a certain moment, the fluid in the conduit with its
mass must be braked or decelerated to zero velocity. At the theoretical limit of instantaneous valve closure and with an incompressible
fluid, the deceleration would be infinitely large, producing a strong
shock at the valve due to the inertia forces at the limit, an infinitely
high pressure shock.

Due to the compressibility of the real fluid, a shock wave is
built up at the valve, travelling through the conduit. The compressibility of the fluid is assisted by the elasticity of the walls of the
conduits. The generation and the velocity of such shock waves will
be examined in this section.

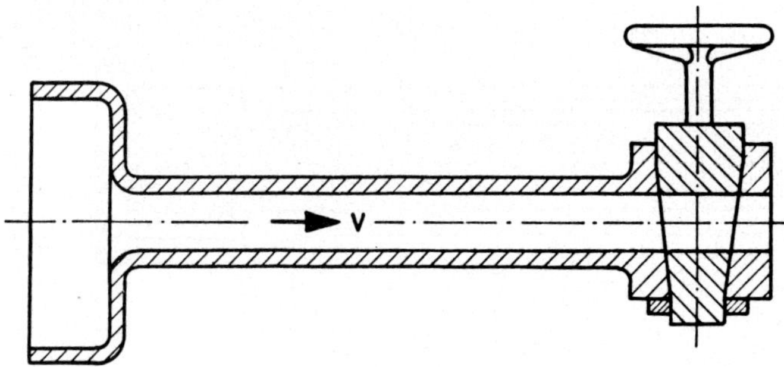


Fig. 72.1: Hydraulic  conduit  with  valve  at  the  right

We use as a basic arrangement the conduit in fig 1, with a large
tank at the left connected by the conduit to the valve at the right.
The mean velocity is determined by the flow and the cross section of
the conduit. We suppose that the flow is selected by external means,
for instance, by variation of the fluid pressure in the tank.

Suddenly closing the valve at the right in fig 1, the fluid must be braked down, producing the following two effects:

1. The fluid is reduced in volume due to the pressure according to the compressibility modulus.

2. The walls of the conduit expand due to the elastic stress produced by the fluid pressure.

The upper part of fig 2 indicates the expansion of the walls under pressure. Beginning from the valve, the fluid is braked down and comes to rest, at increased pressure, layer by layer. The braking advances from right to left in the form of a shock wave. The velocity of propagation will be calculated below, while the steepness of increase depends on the closing time of the valve. The pressure development is shown in the upper part of fig 2 for different instants. It should first be noted that the shock wave arrives at the left end of the conduit of the length L after a time of

$$t = \frac{L}{a} \tag{72-1}$$

where a = velocity of propagation of the shock wave.

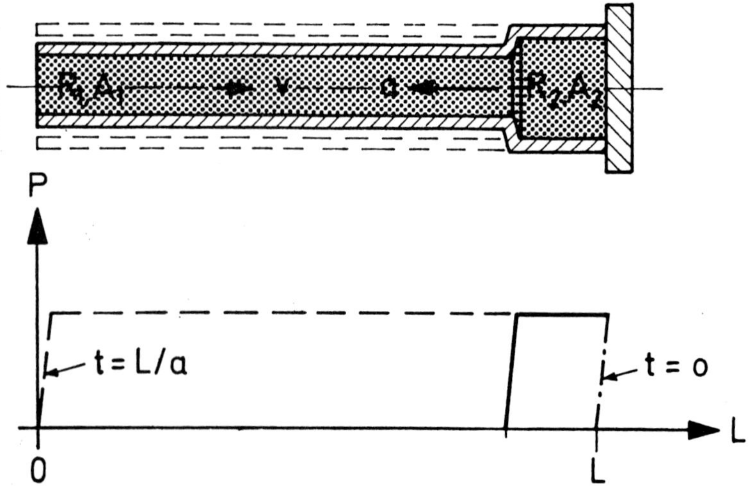


**Fig. 72.2:** Increase of diameter of the conduit and pressure development of the shock wave

Now considering the situation of the conduit at the moment the shock wave reaches its left end, the following changes have taken place since the closure of the valve:

1. The fluid has been completely decelerated and is now at rest.

2. The mass density of the fluid has increased, due to the pressure rise, depending on its compressibility.

3. The walls of the conduit are under elastic stress; it has increased its diameter and, consequently, its volume.

The mass density of the fluid is given by equation 12–1 with the compressibility modulus B as follows

$$\rho_2 = \rho + \frac{p}{B} \tag{72–2}$$

where $\rho$ denotes the mass density before the valve is closed.

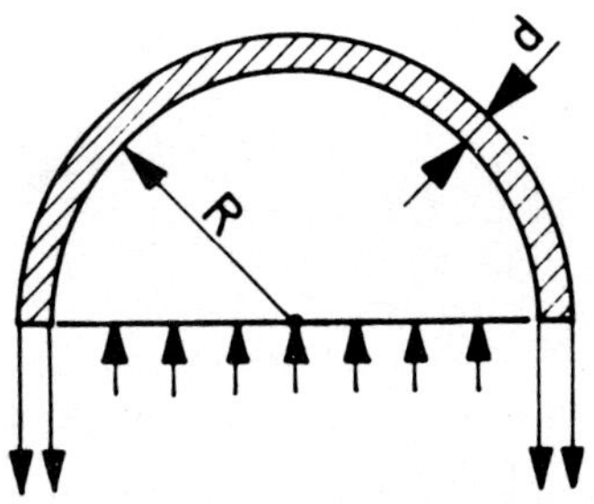


Fig. 72.3: Cross-section of the conduit for calculation of the deformation energy

Fig 3 represents an enlarged cross section of the conduit with the radius R and the wall thickness e. The tangential (also called hoop) stress $\sigma$ is determined by the familiar formula for thin tubes and, at the same time, the elastic deformation or strain $\epsilon$ by Hooke's law with Young's modulus E

$$\sigma = p\,\frac{R}{e} \qquad\qquad \epsilon = \frac{\sigma}{E} = \frac{p}{E}\,\frac{R}{e} \tag{72–3}$$

The strain or relative expansion of the wall material is always very small compared to one. This allows an approximate determination of the increase of the cross-section of the conduit with an increase of the internal radius of $R_2 = R_1 (1 + \epsilon)$

$$A_2 = \pi R_2^2 = \pi R_1^2 (1 + \epsilon)^2 \approx A_1 (1 + 2\epsilon) \qquad (72{-}4)$$

In the last form of equation 4, we have dissolved the parenthesis and neglected the quadratic term.

The potential energy of the liquid after deceleration is composed of:

1. The compressibility energy due to the fluid.

2. The energy contained in the deformation of the walls.

We determine the compressibility energy by recalling from section 12 that an applied pressure produces a proportional reduction of volume. The energy is therefore — as with a spring — equal to half the product of the volume decrease and of the applied pressure (see section 12)

$$E_{comp} = \tfrac{1}{2} V p = \tfrac{1}{2} \frac{p^2}{B} L A \qquad (72{-}5)$$

Strictly, we should have used the cross section $A_2$ of the expanded conduit, but this small difference can be neglected.

The expansion of the wall is also a linear process. The energy is, therefore, equal to half the product of the increase of the circumference of the wall and of the force exerted by the fluid on each part of the circumference. Since this force is equal to pRL (see fig 3) and since the circumference is increased as given by equation 3, we obtain

$$E_{def} = \tfrac{1}{2} 2 \pi R \epsilon \, p R L = \frac{R \, p^2}{e \, E} A L \qquad (72{-}6)$$

We see from formula 6 that the deformation energy increases with softer material or with thinner walls. Any reinforcement of the walls, either by larger wall thickness or by external rings, reduces the energy of deformation.

Apart from the tangential forces treated here, there are still the longitudinal stresses. The consequent potential energy can be calculated similarly; it has the same form as given by equation 6, but only ¼ of its magnitude. It could be taken into account by introducing an equivalent Young's modulus of about 20% less. This correction and also the change of the length of the tube due to the circumferential stress and the Poisson's modulus will be neglected here. It should also be noted that the longitudinal stress produces a prolongation of the conduit and the corresponding potential energy can be absorbed only if the conduit is free to expand. This is the case with free ends of the conduit but not with tubes that are fixed between heavy machines. With curved tubes, one can generally count on free expansion due to their low resistance to bending.

Furthermore, it should be noted that both the compressibility energy and the deformation energy of the walls are proportional to the length of the conduit, due to the assumed symmetry in the longitudinal direction.

The velocity of propagation of the shock wave and its amplitude will be determined by an energy and mass balance. With instantaneous closure of the valve, these two variables are constant, since no fluid is lost and since the entire kinetic energy of the fluid flow is converted into potential energy by deceleration.

### Mass Balance

The mass of the fluid in the conduit at the moment when the shock wave has reached its left end is composed of the mass that was already in when the valve closed and the mass that entered during the travel time of the shock wave. This leads to the following mass balance

$$\rho_1 \, A_1 \, [L + v\,t] \; = \; \rho_2 \, A_2 \, L \tag{72-7}$$

Inserting equations 1 to 4 and neglecting products of the small variables $p/B$ and $\epsilon$, we obtain

$$\rho_1 \, A_1 \, L \left[ 1 + \frac{v}{a} \right] = \rho_1 \, A_1 \, L \left[ 1 + \frac{p}{B} + 2\,\frac{p\,R}{e\,E} \right] \tag{72-8}$$

In order to simplify equation 8, we introduce an *apparent compressibility modulus,* including the deformation of the walls equal to the normal compressibility modulus of the fluid with infinitely stiff walls, i.e. with $E = \infty$.

$$B_2 = B \Big/ \left[ 1 + 2 \frac{B\ R}{E\ e} \right] \qquad (72\text{--}9)$$

Combining equations 8 and 9 simplifies the mass balance as follows

$$\frac{v}{a} = \frac{p}{B_2} \qquad (72\text{--}10)$$

Equation 10 gives therefore, a relation between the flow velocity v, the propagation velocity of the shock wave a and the pressure rise p. The expression $p/B_2$ is generally much smaller than one; in the case of infinitely stiff walls only about 1%, with a shock amplitude of 150 bar.

### Energy Balance

The kinetic energy of the fluid is given by $mv^2/2$ with the mass of the left side of equation 7 or 8. The energy balance becomes, using the potential energies of equations 5 and 6

$$\rho\ A\ L\ \left[ 1 + \frac{v}{a} \right]\ \frac{v^2}{2} = \tfrac{1}{2}\ \frac{p^2}{B}\ A\ L + \frac{R}{e}\ \frac{p^2}{B}\ A\ L \qquad (72\text{--}11)$$

The expression v/a in the parenthesis of equation 1 can be neglected compared to 1. Introducing $B_2$ from equation 9, we obtain

$$\rho\ v^2 = \frac{p}{B_2} \qquad (72\text{--}12)$$

Equation 12 is the result of our energy balance. Combining now both balances 10 and 12 gives us the velocity of propagation and the pressure amplitude of the shock waves singly

$$a = \sqrt{B_2\ /\rho} \qquad (72\text{--}13)$$

$$p = \rho\ a\ v \qquad (72\text{--}14)$$

The velocity of propagation of shock waves is equal to the velocity of sound, known from elementary physics.

It is a notable feature of equations 13 and 14 that both pressures and propagation velocity of the shock wave are independent of the length of the conduit because all energies are proportional thereto. In conduits with constant cross section, no part of the length has any special properties and, consequently, the above derivation is valid also for very short conduits. The only condition is a rapid closing of the valve, while the effect of slow valve closure will be treated in section 73.

**Numerical Example of Joukowsky's Formulae 13 and 14**

Through a steel tube ($E = 2 \cdot 10^6$ bar) with 20 mm internal diameter, 1 meter length and walls of 2 mm, oil flows with a mean velocity of $v = 10$ m/sec. What is the pressure amplitude and the velocity of the shock wave on sudden valve closing?

Using the values

$$B = 1.5 \cdot 10^4 \text{ bar} \qquad R = 1 \text{ cm} \qquad e = 0.2 \text{ cm,}$$

we calculate first the apparent compression modulus by equation 9

$$B_2 = 1.5 \cdot 10^4 \text{ bar} \left/ \left[ 1 + \frac{2 \cdot 1.5 \cdot 10^4 \text{ bar} \cdot 1 \text{ cm}}{2 \cdot 10^6 \text{ bar} \cdot 0.2 \text{ cm}} \right] \right.$$

$$B_2 = \frac{1.5 \cdot 10^4}{1.075} \text{ bar} = 1.4 \cdot 10^4 \text{ bar}$$

The elasticity of the walls reduces the natural compressibility modulus of the fluid by only 7%. The propagation velocity becomes

$$a^2 = \frac{1.4 \cdot 10^4 \text{ daN/cm}^2}{0.9 \cdot 10^{-3} \text{ kg/cm}^3} = 1.55 \cdot 10^{10} \frac{\text{cm}^2}{\text{sec}^2}$$

where we use 1 daN $= 10^3$ kg$\frac{\text{cm}}{\text{sec}^2}$ (compare the numerical example of section 11). Furthermore,

$$a = 1.25 \cdot 10^5 \frac{\text{cm}}{\text{sec}} = 1250 \frac{\text{m}}{\text{sec}}$$

With infinitely stiff walls, we would have obtained

$$a = \sqrt{\frac{B}{B_2}} \cdot 1\,250\ \frac{m}{sec} = \frac{1.4}{1.5}\ 1\,250\ \frac{m}{sec} = 1\,300\ \frac{m}{sec}$$

The velocity of propagation or velocity of sound in oil is much higher than in air, where it amounts to 330 m/sec whilst in steel it is as high as 5 500 m/sec.

The pressure of the shock wave becomes by equation 14

$$p = 0.9 \cdot 10^{-3}\ \frac{kg}{cm^3} \cdot 1\,250 \cdot 10^5\ \frac{cm}{sec} \cdot 10^3\ \frac{cm}{sec}$$

$$p = 1.14 \cdot 10^5\ \frac{kg\ cm}{sec^2} = 1.14 \cdot 10^5 \cdot 10^{-3}\ \frac{daN}{cm^2} = 114\ bar$$

The pressure in the shock wave is astonishingly by large since it would amount to 120 bar with stiff walls. For rapid calculations, we can set up the following practical formula

$$p = 120\ bar\ \frac{v}{10\ m/sec} \qquad\qquad (72\text{--}14a)$$

With a velocity of 1 250 meter/sec, the shock wave needs only 0.8 m/sec to run through the conduit of 1 m length. After this short time, the entire conduit is under pressure, i.e. in the state for which we have set up our energy and mass balances. The valve must be closed very rapidly in order to produce a steep flank of the shock wave. In practice, slower valve closings are more frequent, and will be treated in section 73.

In order to determine the velocity of sound in rubber tubes, it is preferable to determine the reduced compressibility modulus not from the properties of the material but by measuring the volume increase under pressure. According to Guillon, B 7, page 156, the volume of usual high pressure rubber hoses increases by 20% under 200 bar. This corresponds to

$$\frac{\Delta V}{V} = \frac{p}{B_2} \qquad 0.2 = \frac{200\ bar}{B_2} \qquad B_2 = 1\,000\ bar$$

The velocity of sound becomes therefore

$$a = 1\,250 \ \frac{m}{sec} \ \sqrt{\frac{B_2}{B}} = 1\,250 \ \frac{m}{sec} \ \sqrt{\frac{1}{15}} = 320\,m/sec$$

Strictly speaking, we should have added the compressibility of the oil itself but this can be neglected here and with still softer rubber hoses. Practically, the entire potential energy is then contained in the deformation of the walls of the hose.

According to J. Faisandier, /FN¹/ low pressure hoses expand by 40% of their cross section (elongation is neglected) under a pressure of only 10 bar. This supplies as above

$$0.40 = \frac{10\ bar}{B_2} \qquad\qquad B_2 = 25\ bar$$

$$a = 1\,250 \ \frac{m}{sec} \ \sqrt{\frac{25}{15\,000}} = 51\ m/sec$$

According to Comolet, /FN²/ the velocity of shock waves in blood vessels is still lower for the same reasons, in the region of 15 m/sec.

Naturally, all these calculations with soft tubes cannot claim great accuracy since the elongation of the tube has been neglected. Also, with such strong expansion of the walls, some of the above approximations become inaccurate.

The relation between wall expansion and pressure can become non-linear with very soft walls. This leads to distortion of the shock waves in contrast to their normal propagation without change of their form in steel tubes.

FN¹ — Private communication to be included in the new, 1970 edition of B4.

FN² — R. Comolet, 'Mecanique des fluides', volume 1, page 204, Masson, 1961.

## Historical Remarks

Equation 14 was derived by Joukowsky (1847–1921) and rederived by Allievi (1856–1941). In 1913, Allievi gave another graphical method that was later completed by L. Bergeron and O. Schnyder. Whilst Allievi was interested in shock waves in water power stations, N. Joukowsky worked on the water supply of Moscow and published as early as 1898 equations 13 and 14. He is also known for his formula for the lift of infinitely large aircraft wings, discovered in 1905, known today as the Kutta-Joukowsky formula. For further indication, consult the very interesting book H. Rouse and S. Ince, 'History of Hydraulics', Dower Publications, 1963.

Our derivation of equations 13 and 14 is similar to that in the book of B. Nekrassov, 'Hydraulics', Moscow, 1966, published in Russian, English and French. Nekrassov states that he has taken his derivation directly from the original work of Joukowsky.

### 73     Form of the Shock Waves

We have derived the pressure amplitude and the velocity of shock waves in section 72 from a balance of energy and masses set up for the moment in which the fluid has come to rest after the closing of the valve. This is not a steady state, since the shock pressure pushes the fluid back into the tank. It produces an expansion wave beginning from the tank and travelling down the conduit. It can also be called the negative pressure shock wave and it travels down with the same velocity in reverse, because it is subject to the same mass and energy balance, as represented in fig 1. After the passage of the expansion wave, the same pressure as before valve closing is re-established, but the liquid flows in reverse, that is back to the tank.

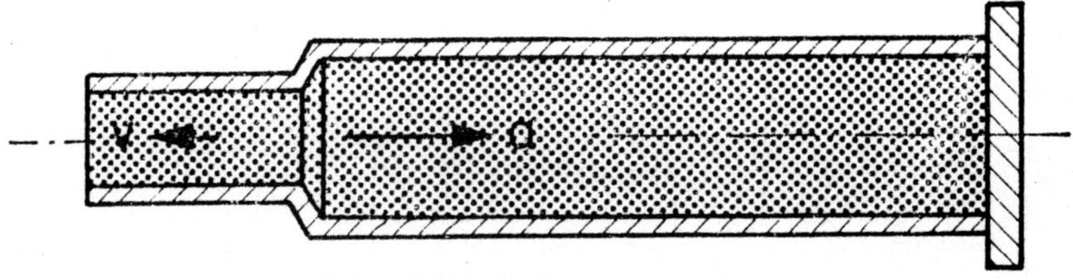

Fig 73.1: Reduction of diameter of the conduit on return
of the shock wave

After a further delay $t = L/a$ the returning expansion wave reaches the valve. There it produces a depression wave, since the closed valve does not supply any fluid, which is the wave with negative pressure according to the equation 72–14. Cavitation will be produced in the conduit if the static pressure at the beginning was smaller than the pressure increase. This would absorb the energy of the wave and can also damage the conduit.

Without cavitation, that is with a weak negative pressure, the depression wave returns to the left, stopping the fluid. With the depression in the conduit, a new shock wave starts at the tank, accelerating the fluid to the right. As soon as it reaches the valve, the

initial state is re-established and the cycle recommences. Of course
the shock waves are gradually damped by the different energy losses
including wall friction, but up to 10 shock waves can be observed in
hydro-electric power plants upon closing a large valve. There the
expansion wave with its depression is especially dangerous, due to
the possible buckling of the conduit under external atmospheric
pressure.

Before examining the situation with slow closing of the valve, it
should be noted that slow means in this context a duration comparable to the travelling time of the shock waves, that is around one
millisecond and above in oil hydraulics. The slow valve closing can
consist of different partial closings; each a little later than the preceding one.   Each of the partial closings produces a shock wave
according to equation 13 and 14. Fig 2 represents a closing consisting
of 4 partial closings evenly distributed in time. The first shock wave
has just arrived at the tank and the other partial shock waves give
a stepwise increase of the pressure. A larger number of partial closings or, at the limit, a continuous closing of the valve, produces a
gradual pressure increase, as shown in fig 2 by an inclined line.

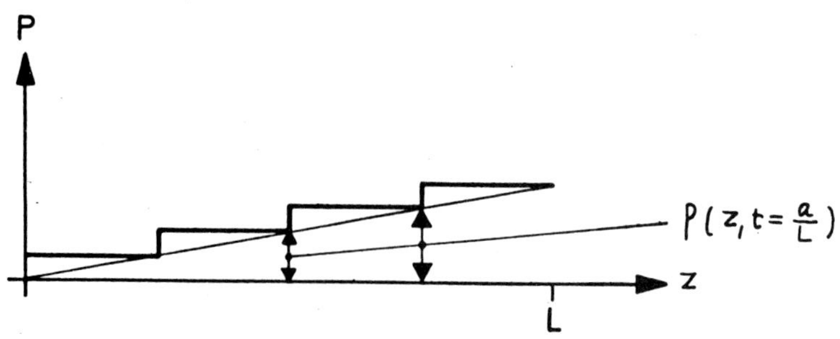


Fig 73.2: Shock  waves  increasing  by  steps  and  at  the
limit  continuously

It should be noted that an even pressure rise is produced only if the
flow velocity in the conduit decreases evenly on closing the valve. Since
a shock wave is immediately built up, increasing the flow through the
valve, the cross-section of the valve must be reduced more rapidly, corresponding to the entire closing time. The power consumed in the valve during
slow closing reduces the amplitude of the pressure shock wave at the
beginning, and does not disturb the energy balance of each single partial
closing.

Fig 3 represents the situation with slow linear pressure rise produced by suitable valve closing. The point broken line shows the pressure after twice the travelling time of the shock wave. It was reflected on the tank and has just reached the valve as a depression wave. The real pressure is equal to the difference between the advancing and the returning wave as indicated by the vertical arrows. It is seen that the actual pressure at the valve is as large as if it were opened suddenly, whilst it decreases linearly towards the tank. If the valve produces a further pressure rise with the same slope, we obtain the pressure development shown by a continuous line in fig 3. Both the advancing and the reflected pressure waves become larger, but their difference remains constant. Therefore the pressure is still the same as at the moment of arrival of the returning shock wave. The result of these considerations is that during slow closure of the valve only a part of the reduction of the flow velocity contributes to the shock wave. This is the part braked down before the returning shock wave again reaches the valve. Consequently we obtain the following equation between the pressure, the closing time $\tau$ and the running time t

$$p = \rho a v \frac{2t}{\tau} = 2\rho \frac{vL}{\tau} \quad \text{if } \tau > \frac{2L}{a} \qquad (73\text{--}1)$$

We would like to note again, that equation 1 is strictly applicable only with linear reduction of flow velocity. Actually it is also used in other cases, e.g. with linear reduction of the valve cross-section.

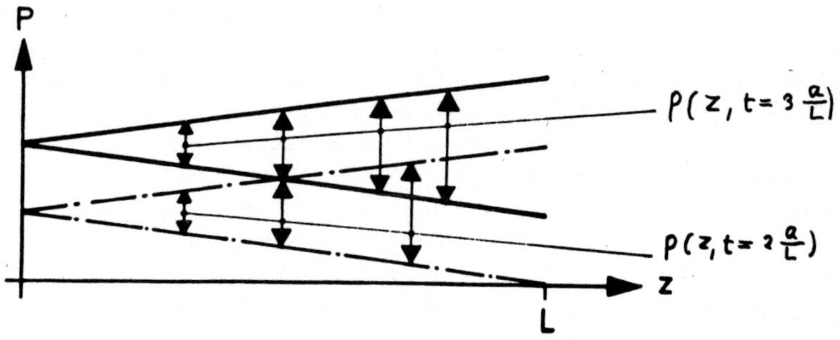


**Fig 73.3: Pressure development with slow closing of the valve**

The pressure amplitude with slow valve closing is independent of the velocity of sound but depends on the length of the conduit. This is represented in fig 4 with the closing time on the x axis and the pressure amplitude at the valve as y axis. As seen in the figure, the pressure amplitude increases with more rapid valve closing until it reaches the maximum given by equation 72 – 13.

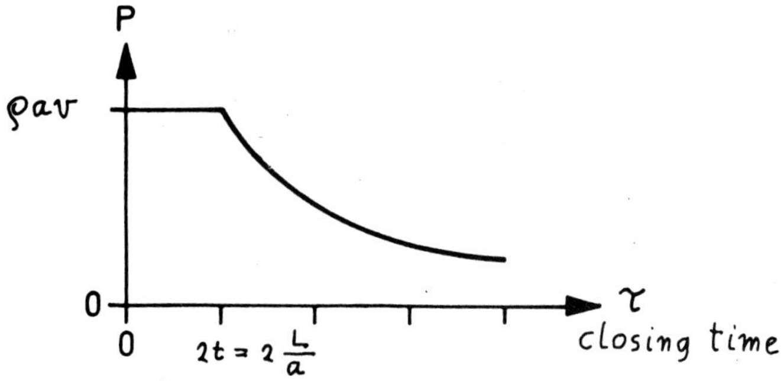


**Fig 73.4:**Maximum    shock    pressure    as    function    of
closing time

Using the data of the numerical example in section 72, the double travelling time, after which the reflected shock wave returns to the valve, is only 1.6 m/sec with a velocity of sound of 1 250 m/sec and a length of 1 meter. Closing the valve in a reasonable time of 16 m/sec, produces only 10% of the pressure calculated above, that is 11 bar (160 psi). The valve faces of axial piston machines close over a rotation angle of about 20 milliradiants (corresponding to 1.2 deg) giving a closing time of 0.125 m sec with a rotation frequency of 160 rad/sec. This rapid closing is one of the reasons why hydrostatic machines are so liable to cause shock waves and noises. Short closing times and therefore possible shocks are also a feature of seating valves, if the beginning pressure wave accelerates the closing itself, as with non-return valves. The above considerations also illustrate the fact, used in section 71, that the steepness of pressure rise is principally responsible for noise production.

Rubber hoses for hydraulics are easily damaged by rapid pressure rise. The specification that a hose can withstand 15 000 bar/sec (225 000 psi/sec) seems very high at first, but corresponds only to a pressure rise of 15 bar (225 psi) in 1 m sec.

We thus see how the shock waves can be reduced by slower closing of valves or generally by slower impedance changes. A further method to reduce pressure shock waves is to install a hydraulic accumulator as near as possible to the valve, or if possible at the position of the highest pressure peak. For calculation purposes one can again set up an energy and mass balance, including the additional potential energy of the fluid in the accumulator. This reduces the pressure amplitude, but this calculation is outside the scope of this book.

As we have seen, the closing of the valve produces a shock wave, where the fluid was first in movement, and comes to rest under higher pressure after the passage of the wave. In a different type of shock wave, the fluid is first at rest under no pressure, when the passage of the wave accelerates and subjects it to pressure. Such waves are produced by sudden acceleration of a piston at the end of a conduit initially filled with fluid at rest. The piston velocity v is imparted to the fluid and runs as a shock wave through the conduit. The relation between pressure amplitude, propagation velocity and flow velocity is again given by the Joukowsky equations 72–13 and 72–14. Such a shock wave can also be caused by the sudden connection of a volume under pressure with a conduit. An example is the connection of the displacement chambers of hydrostatic machines with the delivery port under higher pressure, as happens if the switching timing is not perfectly adjusted.

Summarising, there are two kinds of shock waves in hydraulic systems:

1. Shock waves where the fluid is initially in motion.

2. Shock waves where the fluid is initially at rest.

Both types of shock waves are closely related and can be converted into each other by a suitable moving observer. As an example, fig 2 contains a shock wave of the first kind for an observer at rest, whilst an observer travelling with the velocity a to the right would find a shock wave of the second kind, since the liquid was for him initially at rest.

The study of shock waves with moving observers has been further developed, especially for hydro-electric power plants, and is known as the method of Schnyder-Bergeron. It is used principally to determine the form and amplitude of the pressure shock waves with different valve movements.

# APPENDIX   1

## Oil Hydraulic Circuit Symbols

As in electronics, so with oil hydraulics and pneumatics, it is necessary to represent components and installations by symbols and circuits. The symbols indicate the basic operation of a device and they are in fact a greatly simplified picture or drawing of it. Since they are easy to draw, they save work in representing circuits — especially larger circuits — on paper.

Oil hydraulic circuit symbols were first used by the US automobile industry. The original norms that are still found in some older European publications, were replaced in 1963 by new, more pictorial representations, sponsored by the National Fluid Power Society of the USA. They were adopted, with some slight modifications, by CETOP (Comité européen des transmissions oléohydrauliques et pneumatiques), represented in the UK by AHEM (Association of Hydraulic Equipment Manufacturers). They are also subject to the German norm DIN 24300, pages 1–6.

We list here a selection of the most important symbols.

Conduit or connection for power or control

Auxiliary connection for leakage

Shaft or rod (mechanical connection)

Hydrostatic pumping machine with constant (fixed) displacement

with one direction of flow

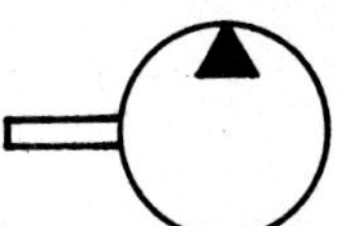

with two directions of flow

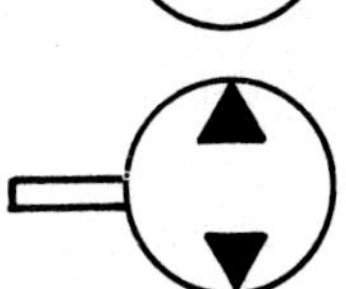

Hydrostatic pumping machine with adjustable displacement

with one direction of flow

with two directions of flow

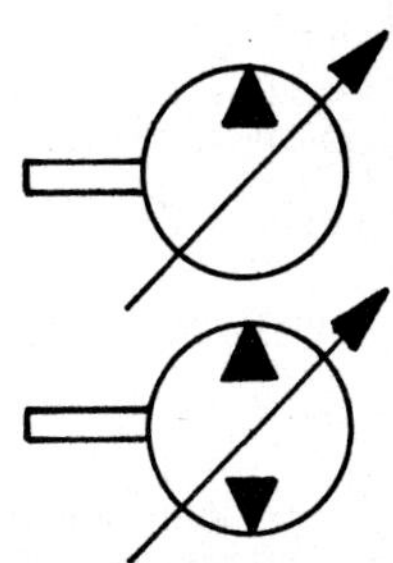

Hydrostatic machine used as motor with fixed displacement

with one direction of flow

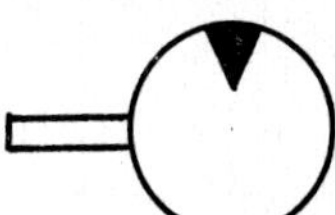

with two directions of flow

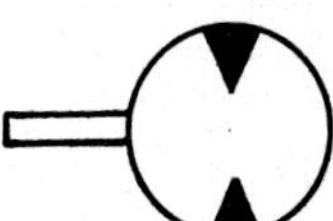

Note : An adjustable displacement is indicated by the oblique arrow.

Actuator or cylinder with thin piston rod

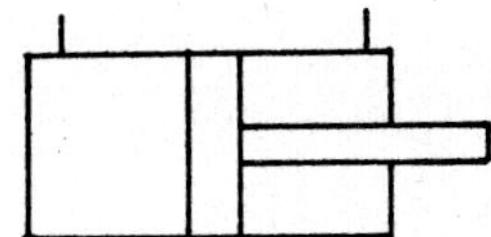

Cylinder with thick piston rod

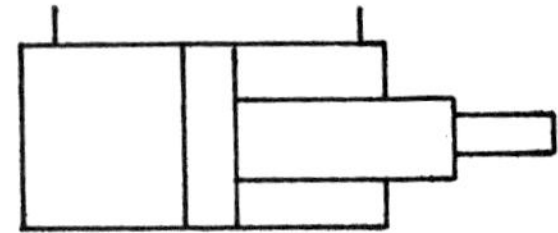

Directional control valves

> Note: Valves with discrete (or fixed) control positions are represented
> by small squares. Each square corresponds to a control position
> and shows the connections made. Such valves are drawn in their
> rest or permanent position and the actuation can be shown at the
> side. Valves with intermediate positions can be indicated by
> double lines around the squares.

Two-way (2/2) valve with electrical
operation and spring reset

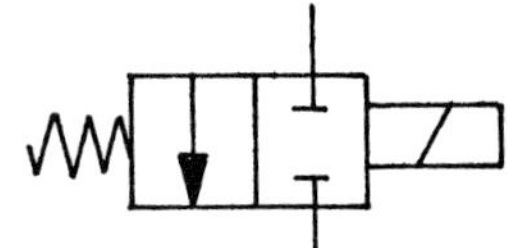

Manually operated 4/3-way valve

1) With blocked flows on both sides
   at rest (blocked centres)

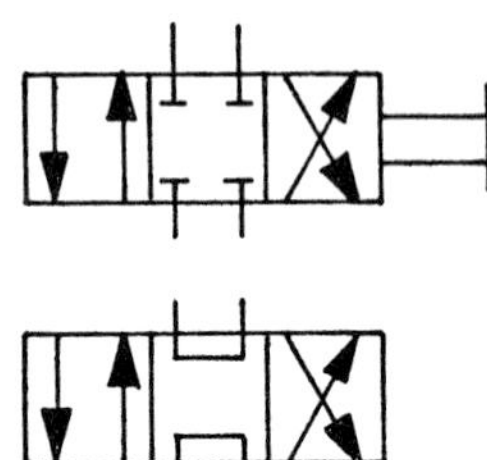

2) With open flows on both sides
   at rest (open centres)

Servo valve, electrically controlled

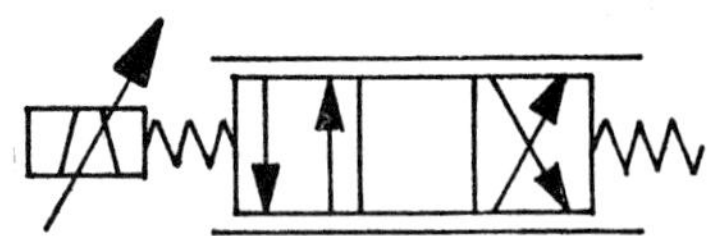

Dipole valves

Non-return valves (in USA check valves)

a) Simple

b) Spring-loaded

c) Piloted by auxiliary connection

Isolating valve, simplified symbol

Throttle valve

a) With preponderantly turbulent pressure drop

b) With preponderantly viscous pressure drop

Note : If adjustable, indicated by oblique arrow

Pressure relief valve

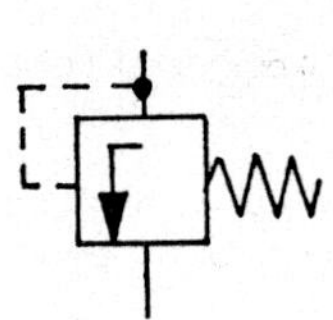

Note : Different positions of the auxiliary
connections indicate any different operation

Flow control valve

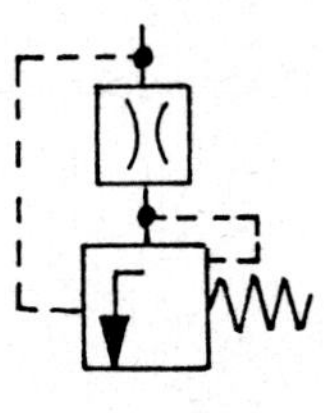

a) Complete symbol

b) Simplified symbol, adjustable

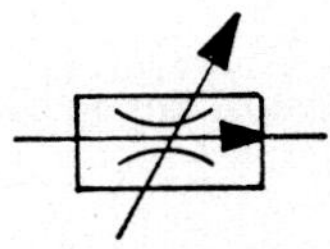

Filter

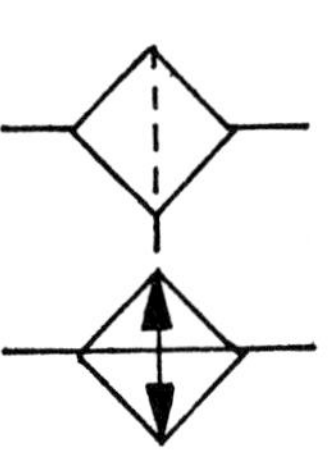

Cooler

Pressure gauge

Thermometer

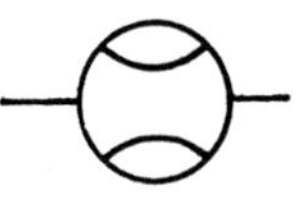

Flow meter

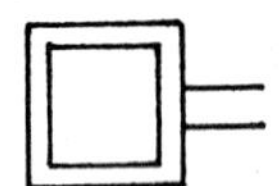

Prime mover,
especially internal combustion engine

Additional remark:

Finally, two especially vivid symbols for servo valves will be shown
here but they have, unfortunately, disappeared from the norms

Double acting servo valve

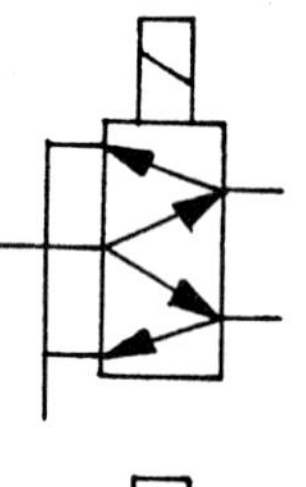

Simple acting servo valve

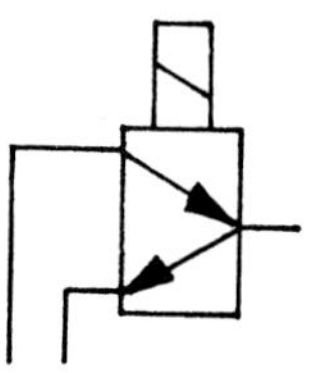

# APPENDIX   2


## Functional Diagrams


In all publications, studies and books there is a great need for graphical communication of ideas. Graphical aids are useful to supplement the text and to avoid too much reliance on mathematical formulae, a common defect of many engineering publications. In this book we have used the following graphical aids :

1) Structural representations, ranging from schematic cross-sections to hydraulic circuits, with the symbols as represented in appendix 1. In such circuits, each symbol represents a real component.

2) Characteristic curves, giving the relations between an input and an output variable, depending sometimes on a third variable or parameter. Such curves are also called simply characteristics.

3) Functional diagrams, representing a grid or net of causes and effects, or of functional relations. Since they are very powerful but perhaps not so familiar, they will be explained further:-

Functional diagrams were first used in control engineering, and are composed of the few elements represented schematically in fig 1.

1) Blocks representing a functional relationship or giving an output variable as a function, sometimes but not always linear, of the input variable.

2) Connections, where a signal flows in the direction of the arrow.

3) Branch points, where a connection splits up without change of the signal.

4) Addition and subtraction points, where the output signal is equal to the sum, or to the difference of, the signals, in the two incoming connections.

5) Blocks subjected to disturbances, as shown by a lateral arrow, representing a functional relation with a parameter.

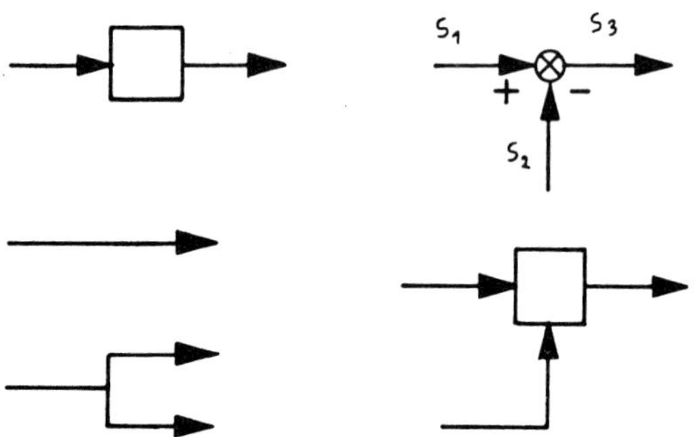


Fig A2.1: Basic elements of functional diagrams

The above conditions are simply conventions justified by their usefulness. The unilateral action of each block is essential where the output is a function of the input only, and independent of what happens to it afterwards.Similarly, in connections, the signal is determined by the input only and not by the output nor by the branch points. Adding one or more branch points to a connection does not change the signal determined solely by the input. Furthermore, each connection contains only one variable.

Real connections contain two variables and almost always have some return effect.For example, a hydraulic conduit contains pressure and flow as variables. If the pressure is considered as the main effect or signal, producing a flow in a load, then this flow has some effect on this pressure, unless supplied by an ideal pressure source. Such effects are indicated in functional diagrams by separate feed-back connections. Furthermore, many components give relations between two pairs of variables, such as hydrostatic machines, which relate rotation frequency and flow on the one hand, and torque and pressure on the other.

The operation of the machine is then represented by two blocks each giving one of these relations with unilateral action, as shown in the example in figs 62–1 and 62–2.

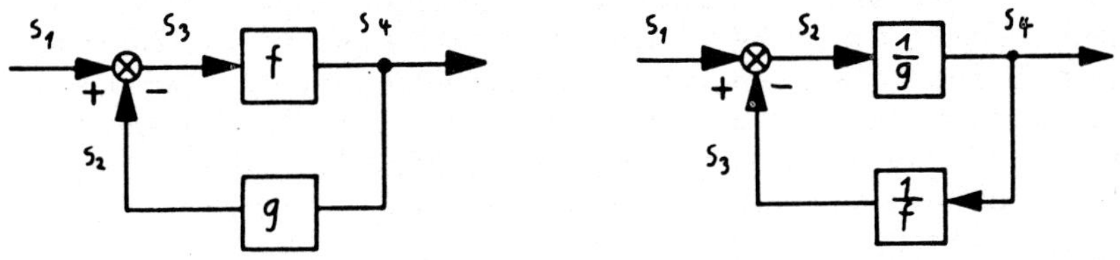

Fig A2.2: Simple functional diagram at the left and its
inversion at the right

Due to the return action of many real components, we frequently obtain feedback loops as known from control engineering. Fig 2 shows at left a basic feedback loop with linear blocks having gains of f and g. It represents the following relations between the variables $S_1$, $S_2$, $S_3$, $S_4$

$$S_3 = S_2 - S_1 \qquad S_4 = f\,S_3 \qquad S_2 = g\,S_4 \qquad (A2–1)$$

It is interesting to know the product of the gains making up the closed loop, the so called *loop gain* equal to f g. The overall transfer function of the functional diagram with closed loop, that is the ratio of the output variable $S_4$ to the input variable $S_1$, can be obtained by elimination of the intermediate signals $S_2$ and $S_3$

$$\frac{S_4}{S_1} = \frac{f}{1 + f\,g} \qquad (A2–2)$$

Equation 2 is the famous formula of Mason and is expressed in words as follows :

*The closed loop transfer function is equal to the gain of the forward part( here f) divided by one plus the loop gain (here f g)./FN¹/*

FN[1] — Similar, but more complicated, formulae for the immediate determination of the transfer function of functional diagrams with several closed loops are known but not often required in oil hydraulic engineering.

Functional diagrams assign the role of a signal to one of the two variables in a real connection, whilst the other variable is ignored or represented as a return (feedback) signal. The relative importance of the return signal is indicated by the loop gain and in particular a loop gain very small compared to one, means that the return signal has negligible effect. The indentification of a variable with a signal, represents really a choice of *causality* and is arbitrary in principle. For example, we can consider the flow as a signal and the pressure as the return effect of a dipole, or alternatively, the pressure as the signal and the flow as a return effect. These choices of causality were called Z and Y representations in Section 14.

Closely related to the choice of causality is the possibility of inversion of closed loops of the functional diagram. Fig 2 represents at right the inversion of the simple functional diagram of the left side. Here $S_2$ becomes the error signal, producing $S_4$ by the block with $1/g$ and the feedback signal $S_3$ by the lower block with the gain $1/f$. Hence the inverted functional diagram is equivalent to the following equations

$$S_2 = S_1 - S_3 \qquad\qquad S_4 = S_2/g \qquad\qquad S_3 = S_4/f \qquad (A2\text{--}3)$$

The loop gain is equal to $1/f\,g$ and the closed loop transfer function becomes

$$\frac{S_4}{S_1} = \frac{1/g}{1 + 1/fg} = \frac{f}{1 + f\,g} \qquad\qquad (A2\text{--}4)$$

As expected, the transfer function is the same in the direct and the inverse functional diagrams, because both are equivalent. Nevertheless the loop gains are reciprocal in both cases. In particular, if the loop gain in the direct case is large compared to one, the loop gain in the inverted functional diagram will be very small compared to one, and the return action may then frequently be ignored.

Some blocks representing real components can be inverted without hesitation, like the relation between torque and pressure of hydrostatic machines. Other blocks, including the relation between displacement setting and the flow of a hydrostatic machine are irreversible. In fact, trying to change the flow with a given $\alpha$ of a hydrostatic machine, produces dangerous pressures and might destroy the machine, but will certainly not influence the displacement setting /FN$^2$/ Nevertheless, in closed loops all blocks may be inversed, if only as a conceptual aid.

Inversion is, therefore, a powerful additional tool and operation with functional diagrams.In practice it allows us to set up causalities from physical intuition or knowledge of the operation, and then to invert one or more loops for easy calculation.

Functional diagrams are also needed for computer simulation. The interaction of variables in the network of causes and effects can be directly programmed on analogue computers, where each variable is represented by a voltage. The amplifiers are such that the associated currents produce only negligible perturbations.

Hydrostatic systems contain many pairs of variables, the product of which is a power, thus showing a certain symmetry. Power is also approximately conserved in many transformations. /FN$^3$/ Any deviation from power conservation is indicated by the efficiencies.

Based on the special properties of power, a new graphical representation, the Bond graphs, have been introduced by Prof.Paynter. They allow systematic construction of functional diagrams and choice of causalities including inversions. For a detailed treatment the reader is referred to :

D.Karnopp and R.Rosenberg — Analysis of Multiport
    Systems — MIT Press 1968.

FN$^2$ — Quite different is the behaviour of a hydrostatic machine equipped with a pressure regulator, where the flow can be selected by a suitable load impedance.

FN$^3$ — Amusing remark from the lecture ' 'Power is quite a special elixir'

# APPENDIX   3

## Dimensions, Variables and Numbers

Many publications are not very clear or use different ideas about the distinction of the variables of physics and engineering, their dimensions and of numbers. Here is the summary of our ideas, as used throughout this book. In our opinion these bring much more clarity into engineering and physics.

The variables, sometimes called quantities in physics, are the fundamental concepts of physics and engineering. In hydrostatic engineering, the most prominent variables are pressure and flow, rotation frequency and torque. Of quite a different nature are the numbers, the traditional object of mathematics. They originated as integers, by counting discrete objects and were soon extended to more complex cases, like negative and non integers (rational, irrational and transcendental) numbers.

The relation between variables in physics and numbers, is set up simply by the introduction of unit variables commonly called *units*. The instantaneous value of each real variable is expressed as a multiple of the unit variable. The multiple, sometimes called measure, is a number. For example, the real pressure is denoted as the product of the measure $p_m$ and the unit of pressures $p_u$

$$p = p_m \, p_u \quad = \text{ say 5 bar } = \quad 73 \text{ psi}$$

If the units expressing a given variable are changed, the measure must be changed also. For example, going from bar to psi, a unit 14.5 times smaller (since 1 bar = 14.5 psi) the measure must be multiplied by 14.5

In all processes and laws of physics of engineering the variables themselves are essential and not the measures, which depend on the arbitrary units. It is therefore desirable to establish the formulae and equations, so that they involve the variables themselves.

This is indeed possible by observing some special rules for calculations with the variables, different from the rules of calculation with numbers. The important property of the variables governing the calculation rules, is that they fall into several different classes, commonly called *dimensions*. Bearing this class distinction in mind, the rules are as follows :

1) Only variables of the same dimensions can be equated, added or subtracted.
2) Variables of different dimensions can be multiplied.

3) Multiplication of a variable by a number, gives a variable of the same dimension.
4) Variables of the Null class (dimensionless variables) are numbers as obtained by division of two variables of the same class.

Rule (1) precludes the addition of two variables of different dimensions, like flow and area in hydraulics. Multiplication /FN$^1$/ of variables of different dimensions is possible by rule (2) and allows to pass the class limits. The properties of variables are illustrated in fig 1, where all variables of a dimension (or class) are contained in each hexagon which carries the symbol usually denoting the class

Each variable of a given dimension can be represented by a small point in a hexagon. Combining two variables of a class by multiplication produces a new representative point in a third hexagon. Furthermore, several multiplications, or generally power products of variables, are possible. On the other hand, polynomials cannot be formed unless the parts are dimensionless, because they comprise additions precluded by rule (1).

According to the above rules, equations always link variables of the same class, like pressure or flow. Dividing such an equation, say between flows, by a suitable reference flow, produces an equation between dimensionless flow ratios.

FN$^1$ — and division, which can be thought of as multiplication with the reciprocal

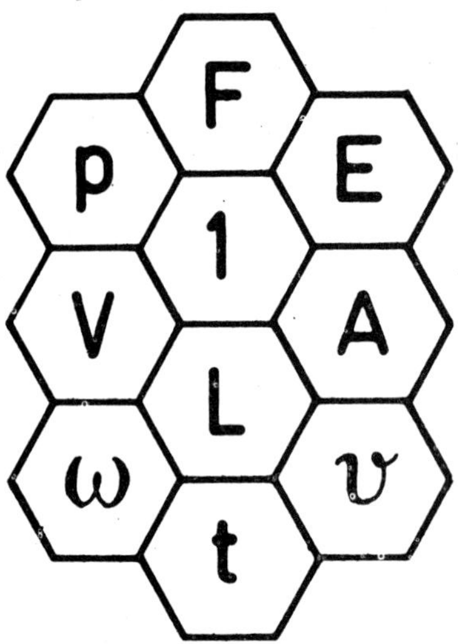

**Fig A3.1:**Representation of the variables of physics and
engineering where all the variables of a dimension or
class are contained in the hexagon carrying the usual
symbols for this class. The central hexagon with the
symbol 1 contains the dimensionless variable

All variables of physics are built from a very few fundamental
ones, in mechanics the three fundamentals length, time and force
/FN[2]/

The building and the properties of the variables are illustrated in
fig 2, with the fundamental variables called $u_1$, $u_2$, $u_3$. They combine
as power product in $q_1$, $q_2$ . . . $q_5$ forming the usual variables of
physics, normally referred to as derived variables. Finally, all
equations between the variables can be reduced to a relation between
the dimensionless $\pi$ products by dividing by a reference variable, as
indicated above. The number of the $\pi$ products is always less than
the number of physical variables, simplifying calculations and results.

FN[2] — Many publications use mass instead of force as a fundamental quantity, a
choice possible in principle but not advisable in practice, in the author's opinion.

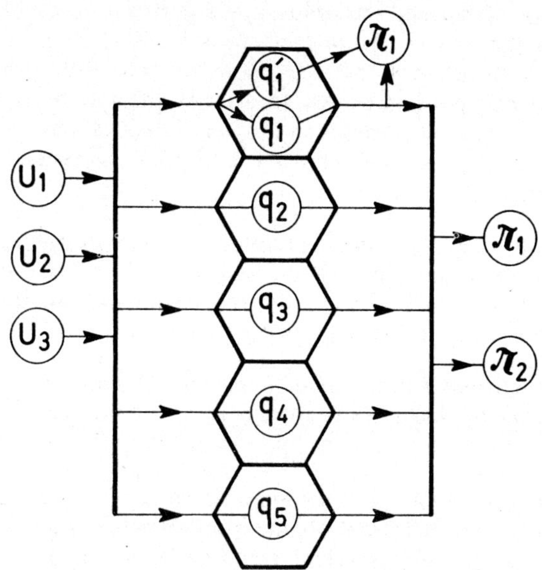

Fig A3.2:Representation of the relations between the
fundamental variables $u_1$, $u_2$, $u_3$, the derived variables
$q_1$ . . . $q_5$ and the dimensionless power products
$\pi_1$ and $\pi_2$

This reduction is called the Pi-theorem /FN[3]/.

More precisely, a general relation between n variables built from
three fundamental variables can be represented by n − 3 dimensionless $\pi$ products.

Examples of the $\pi$ products are the Reynold's number, and the
expression like $\omega\mu/p$ and $p/B$ in Sections 31 and 36. Whilst the units
are arbitrary in principle, for derived variables it is convenient to
use units obtained by the combination of the units of the fundamental
variables. For example, for the unit of pressure psi, one uses the
pressure exerted by a pound force on a square with one inch length
and width. Such a choice permits the insertion of the units in the
formula and to divide them out, an operation which complies with the
above rules. This dimensional check helps to discover errors.

FN[3] — Jean Thoma : Physical Quantities and Dimensional Analysis, Bulletin of the
Mechanical Engineering Education, September 1968.

The units of derived variables are sometimes denoted by a combination of the names of the fundamental units, such as pound per square inch or by new names, such as bar. With more involved variables it is often convenient to express them as a combination of more familar derived variables. For example, we have denoted the unit of viscosity not by daN sec/cm$^2$, but by barsec, easier to remember and to pronounce.

Functional diagrams have variables in their connections and the blocks with their gains can relate variables of different dimensions. The return signal must have the same dimension as the input, by rule (1). This leads to the important conclusion :

*The loop gain of functional diagrams is dimensionless.* Only this fact allows us to compare the loop gain to the number 1, as done in A2.

An interesting application of the properties of variables of physics is the derivation of practical engineering formulae. They are obtained by division of the real physical equation by an equation between a suitably chosen reference variable.For example,consider the dynamic pressure

$$P_{dyn} = \rho \, \frac{v^2}{2}$$

With the reference values $v_r = 10$ m/sec $\rho_r = 10^3$ kg/m$^3$ (of water)

$$Pr = \frac{1}{2} \, 10^3 \, \frac{kg}{m^3} \, 100 \, \frac{m^2}{sec^2} = 0.5 \text{ bar}$$

Since 1 bar $= 10^5 \, \dfrac{N}{m^2} = 10^5 \, \dfrac{kg}{m \, sec^2}$

Now, dividing the dynamic pressure by the relation between the reference values

$$\frac{P_{dyn}}{0.5 \text{ bar}} = \frac{\rho}{10^3 \text{ kg/m}^3} \left[ \frac{v}{10 \text{ m/sec}} \right]^2 \qquad (A3-1)$$

or with $\rho = 0.9 \; 10^3$ kg/m$^3$ for hydraulic oil

$$ p_{dyn} = 0.45 \text{ bar} \left[ \frac{v}{10 \text{ m/sec}} \right] \qquad\qquad \text{(A3-2)} $$

which justifies the rule of Section 13: *The dynamic pressure belonging to a velocity of 10 m/sec is equal to 0.45 bar, increasing with the square of the velocity.*

The strength and deflection of coil springs is a further interesting application. Considering torsional stresses only, the load and the deflection are given as follows : /FN[4]/

$$ F = \frac{\pi}{16} \, d^2 \, \frac{d}{D} \, \tau \qquad\qquad k = \frac{1}{8} \, \frac{d}{z} \, \frac{d^3}{D^3} \, G \qquad \text{(A3-3)} $$

With the reference values $d$ = wire dia ( = 1mm),
$D$ = coil dia (10mm), $\tau$ = shear stress ( = 25 hebar),
$G$ = modulus of elasticity in shear ( = 800 hebar),
$z$ = number of turns ( = 1), we obtain

$$ F_r = 1 \text{ daN}, \qquad\qquad k_r = 1 \text{ daN/mm} $$

Now dividing equation 3 by the relation between the reference values

$$ F = 1 \text{ daN} \, \frac{d}{1 \text{ mm}} \, \frac{10d}{D} \, \frac{\tau}{25 \text{ hebar}} $$

$$ \text{(A3-4)} $$

$$ k = \frac{1 \text{ daN}}{\text{mm}} \, \frac{d}{1 \text{mm}} \, \left[ \frac{10d}{D} \right]^3 \, \frac{1}{z} $$

Equation 4 can be expressed in words : *A coil spring with a diameter ratio of 10 and a shear stress of 25 hebar has an admissible load of 1 daN, and each turn deflects by 1 mm with this load.* /FN[5]/

In conclusion, it is important to keep in mind the dimension or class of each variable,and clearly distinguish them from the numbers. Avoiding any confusion here, gives much more power to reasoning, analysis and design.

FN[4] — M.F.Spotts, Mechanical Design Analyses, Prentice Hall, 1964.

FN[5] — As communicated to the author by Prof.Viersma,Technical University of Delft, Holland.

# BIBLIOGRAPHY

This list contains several books that are useful for further study and to which frequent reference is made in this book. Less important or rarely used references are given as footnotes of the text.


1) J.Blackburn and others, 'Fluid Power Control', Wiley, 1960, (also in French and German). Comprehensive treatment of servo controls, grown out of research at the University MIT, USA.

2) A.Dürr and O.Wachter, 'Hydraulik in Werkzeugmaschinen', 6th edition, 1968, Carl Hanser Verlag. (In German). A beautiful book about industrial oil hydraulics, with many design examples.

3) W.Ernst, 'Oil Hydraulic Power', 2nd edition, McGraw Hill, 1960. Industrial hydraulics in American practice.

4) J.Faisandier, 'Les mécanismes hydrauliques', 2nd edition, Dunod, 1962. (In French). Oil hydraulics and components with many detailed calculations, especially for aircraft controls. 3rd edition expected 1970.

5) F. and D.Findeisen, 'Oelhydraulik in Theorie und Anwendung', 2nd edition, 1968, Schweizer Verlagshaus, Zürich. (In German). Introduction to oil hydraulics, with many calculations.

6) D.Fuller. 'Theory and Practice of Lubrication for Engineers', Wiley, 1956 (Also in German). Textbook of hydrostatic and hydrodynamic lubrication, especially for large machines.

7) M.Guillon, 'Hydraulic Servosystems', Butterworth, 1960 (original in French, also in German, Russian and Polish). An excellent textbook about hydraulic servocontrols with many component calculations, especially about electro—hydraulic servo—valves,

8) R.Hadeckel, 'Displacement Pumps and Motors', Pitman, 1951. Detailed geometric treatment of a variety of mechanisms that can be used as displacement machines.

9) J.C.Hunsacker and B.G.Rightmire, 'Engineering Fluid Mechanics', McGraw Hill, 1947. Fluid mechanics from the point of view of the engineer, with applications of the pi-theorem.

10) H.E.Merrit, 'Hydraulic Control Systems', J.Wiley, 1967. Servocontrols in industry,with detailed control engineering treatment.

11) R.Pfab, 'Getriebe-Atlas', A.G.T.Verlag, 1967. (In German). Survey of designs of hydrostatic transmissions taken mainly from patent specifications.

12) F.Reuleaux, 'The Kinematics of Machinery', (original German, English edition 1876, reprinted by Dover Publications in 1963). Fundamental work about the kinematics of displacement machines, at its time steam motors, but interesting still in the 1970's.

13) W.Stiess, 'Pumpen-Atlas', volume 1, 'Displacement Machines', A.G.T.Verlag, 1967 (In German). Survey of displacement machines for oil hydraulics and for non-lubricating fluids.

14) J.Thoma 'Hydrostatic Power Transmission' Trade and Technical Press, 1964 (also in French, German, Italian and Spanish). A more detailed treatment than in the present book about the theory and application of hydrostatic transmissions in industry.

15) W.Wilson, 'Positive Displacement Pumps and Motors', Pitman, 1950. Theory and experiments with hydrostatic machines containing the original mathematical model of Wilson.

## Additional Titles

The following titles are recommended for further study although they are not frequently referred to in this book:

16) N.S.Grassan and J.W.Powell (editors), 'Gas Lubricated Bearings', Butterworth 1964. Excellent report, grown from a lecture programme at the University of Southampton about theory and application of compressible lubrication, with design examples of gas bearings.

17) D.Karnopp and R.Rosenberg, 'Analysis and Simulation of Multiport Systems', MIT Press, 1968. Fundamental work about theory and application of bond graphs, recommended as introduction to this very powerful and elegant tool for analysis and design.

18) E.Muijderman, 'Spiral Groove Bearings', Philip's Technical Library, 1966. Complete treatment of hydrodynamic bearings with pressure generation by spiral grooves.

19) O.Pinkus and B.Sternlicht, 'Theory of Hydrodynamic Lubrication', McGraw Hill, 1961. Detailed theoretical treatment including effects of variable temperature and of compressibility.

# CONTRIBUTORS OF FIGURES

As denoted by a short-word in the captions, some figures were contributed by several firms. This list contains the full address of these firms listed following the short-word in the caption.

ATE
Alfred Teves GmbH, D 6 Frankfurt
(figure supplied by Bremstechnik AG, Bern)

Cattermole
Cattermole Hydraulics Ltd.,
Astwood Bank, near Redditch, England.

Eckerle
Otto Eckerle, Postfach 40, D 7502 Malsch.

Flygmotor
Flygmotor AB, Trollhättan, Schweden.

Galdabini
C. Galdabini SpA, C.p. 199, I 21013 Gallarate.

Herion
Herion Werke KG,
Postfach 2970, D 7 Stuttgart 1.

Hydroméca
Hydroméca,
83, bd de Charonne, F 75 Paris 11.

Indramat
Indramat GmbH,
Postfach 409, D 8770 Lohr.

Regeltechnik · Regeltechnik,
                Dipl. Ing. F. Landwehr
                Postfach, D 799 Friedrichshafen.

Ruston          Ruston Hydraulic Motor Division,
                P.O. Box 25, Lincoln, England.

S.I.G.          Schweizerische Industrie-Gesellschaft,
                CH 8212 Neuhausen.

Vickers Sperry  Vickers Sperry Rand Ltd,
                P.O. Box 4, Havant, Hampshire, England.

VSG             Vickers-Armstrong Div.,
                P.O.Box 8, Swindon, Wiltshire, England.

Wandfluh        Wandfluh AG, CH 3716 Kandergrund.

# GLOSSARY OF ENGINEERING TERMS

Oil hydraulic engineering is essentially an international technique with the same problems arising all over the world. We include here a small glossary of important engineering terms, in order to facilitate communication of ideas between different language regions.

The terminology is far from unified, especially in patent specifications. This glossary contains the terms preferred by the author, while some other terms, used in the literature are added in round brackets. The square brackets contain different meanings or alternatives.

| English | French | German |
|---|---|---|
| Chapter 1 | | |
| hydrostatic machines | machines hydrostatiques | hydrostatische Maschinen |
| (hydrostatic units) | (volumétriques) | (hydrostatische Einheiten) |
| hydrostatic components | composantes hydrostatiques | hydrostatische Bauteile |
| volume flow, | débit (de volume) | Volumenstrom, Oelstrom |
| (flow rate) | | (Fördermenge) |
| pressure | pression | Druck |
| torque | couple (moment tournant) | Drehmoment |
| rotation frequency | fréquence (vitesse) de | Drehzahl, Drehfrequenz |
| (rotative speed) | rotation | |

| English | French | German |
| --- | --- | --- |
| Chapter 1 (continued) | | |
| power | puissance | Leistung |
| viscosity | viscosité | Viskosität (Zähigkeit) |
| dynamic pressure | pression dynamique | Staudruck |
| secant, tangent, impedance, admittance | impédance, admittance de sécante, de tangente | Sekanten-, Tangenten-impedanz, -admittanz |
| Chapter 2 | | |
| gap formula | formule des fentes | Spaltformel |
| gap height | hauteur de fente | Spalthöhe |
| spherical piston | piston sphérique | Kugelkolben |
| slipper (pad) | patin | Gleitschuh |
| load ratio | facteur de charge | Belastungsgrad |
| balancing factor | facteur de décharge | Entlastungsgrad |
| sealing lip | lèvres de contact | Dichtlippe |
| overbalanced bearing | coussinet surbalancé | überbalanciertes Lager |
| underbalanced bearing | coussinet sousbalancé | unterbalanciertes Lager |
| tilting moment | couple de basculement | Kippmoment |
| geometrical [thermal] wedge | coin d'huile géométrique [thermique] | geometrischer [thermischer] Schmierkeil |

Chapter 3

| | | |
|---|---|---|
| displacement chamber (volume) | volume de déplacement | Verdrängerraum |
| displacement | cylindrée | Verdrängervolumen |
| positive distribution | distribution forcée | zwangsläufige Steuerung |
| slip (rotation) frequency | vitesse de glissement | Schlupfdrehzahl |
| loss pressure | pression de perte | Verlustdruck |
| displacement setting (inclination ,tilting) | réglage volumétrique (inclinaison) | Volumeneinstellung (Schwenkung) |
| tilting angle | angle d'inclinaison | Schwenkwinkel |
| [swash plate] axial piston machine | machine à pistons axiaux [à plateau inclinable] | Axialkolbenmaschine [mit Schiefscheibe] |
| tilting head machine (bent axis machine — Thoma type machine) | machine à axe coudée | Axialkolbenmaschine mit mit Schwenkkopf |
| valve faces (timing faces) | surfaces de distribution | Steuerflächen |
| kidney (timing) ports | lumières (ouvertures) de distribution | Steueröffnungen |
| metering throttle | étranglement de dosage | Drosselstelle |
| drive flange | plateau de l'arbre | Triebflansch |
| gear machine (pump) | machine (pompe) d'engrenage | Zahnradmaschine |
| vane machine | machine à palettes | Flügelmaschine |
| characteristics | courbes caractéristiques | Kennlinien |

| English | French | German |
|---|---|---|
| **Chapter 4** | | |
| throttle valve | valve d'étranglement | Drosselventil |
| pressure relief valve | régulateur de pression (clapet de surpression) | Druckregelventil (Druckbegrenzungsventil) |
| flow control valve | régulateur de débit | Stromregelventil |
| directional control valve | distributeur | Wegeventil |
| piloted valve | valve pilotée | vorgesteuertes Ventil |
| non return valve (check valve) | clapet anti-retour | Rückschlagventil |
| servo valve | servovalve (servo-distributeur) | Servoventil |
| accumulator | accumulateur | Druckspeicher (Akkumulator) |
| **Chapter 5** | | |
| pressure [power] regulator | régulateur de pression [de puissance] | Druck-[Leistungs-] regler |
| **Chapter 6** | | |
| primary | primaire | Primarteil |
| secondary | secondaire | Sekundarteil |

Chapter 6 (continued)

| | | |
|---|---|---|
| priming (boost) pressure | pression d'alimentation | Speisedruck |
| apparent (corner) power | puissance apparente | Scheinleistung |
| torque multiplication | multiplication de couple | Drehmomentverstärkung |
| power split transmission | transmission à super-position de puissance | Getriebe mit Leistungs-verzweigung (Ueberlagerungsgetriebe) |

# INDEX

330

# INDEX TO ADVERTISERS

# Pan-Sec Hydraulic valve panels cut costs

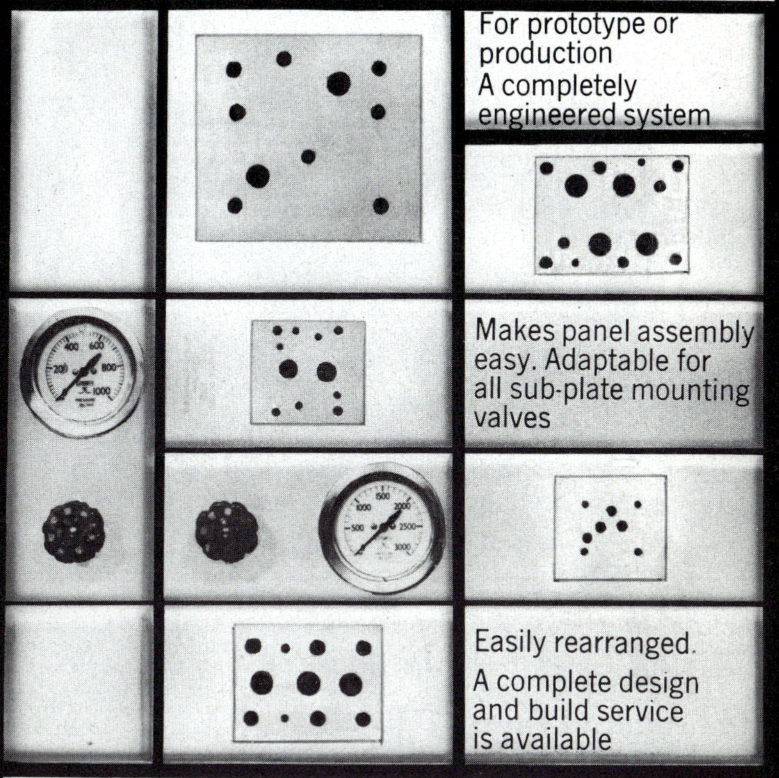

For prototype or production
A completely engineered system

Makes panel assembly easy. Adaptable for all sub-plate mounting valves

Easily rearranged.
A complete design and build service is available

For details contact:
Engineers (Sutton) Limited
Kingsmill Lane, South Nutfield,
Redhill, Surrey RH1 5ND

STD: 073-782 Nutfield Ridge 2471   Telex: 261934

# UNIQUE

## a new range of Fairey filter elements giving 1 micron filtration at up to 100 gallons a minute

Fairey introduce a filter capable of fine filtration *and* a very high flow rate. It has a filter element which will remove *all* particles larger than 3 microns, 98% of all particles larger than 1 micron and filter out the ultra-fine silt which contaminates fine clearance components like servo valves. Add the ability to accept flows up to 100 gallons per minute and you have the Fairey F.100 – a truly unique filter element.

**Fairey sets the pace in filtration**
The F.100 is the latest of a whole series of new product developments from Fairey. We started with the push-on element. Then came the reverse flow filter. The latest is the F.100 and there's more to come. We'll keep you informed of all the new developments as they happen. Meanwhile, write and find out about the Fairey F.100. It's unique.

## Fairey Filtration Ltd

Cranford Lane, Heston, Hounslow, Middx.
Telephone: 01-573 7777. Telex: 22230

# A comprehensive range of hydrostatic transmission units

### FRASER —DUSTERLOH

FRASER-DUSTERLOH low speed, high torque hydraulic motors with overall efficiencies of up to 94% are available in 16 sizes. Displacements range from 37 cm³/rev. to 16,500 cm³/rev. with equivalent maximum torques of 5.45 kg.m. to 3,940 kg.m. Speed ranges cover 5 to 1,500 rpm.

### FRASER FR PUMP/MOTOR RANGE

FRASER FR PUMP/MOTOR UNITS serve either as a pump or a motor without modification. They are made in 12 sizes — as pumps with outputs from 6.81 1/min. to 136 1/min. at 1,500 rpm; as motors, continuous torques range from 1.4 kg.m. to 24.2 kg.m. at speeds of up to 3,000 rpm. All units are available for either foot or flange mounting.

Write or telephone for detailed literature

# ANDREW FRASER & CO. LTD.

Hampton Road West, Hanworth, Feltham, Middlesex.

Tel: 01-898 0961          Telex: 261685